continued on back

Planning and Analysis
of Observational Studies

Planning and Analysis of Observational Studies

WILLIAM G. COCHRAN

Edited by

LINCOLN E. MOSES

and

FREDERICK MOSTELLER

John Wiley & Sons

New York • Chichester • Brisbane • Toronto • Singapore

Library of Congress Cataloging in Publication Data:

Cochran, William Gemmell, 1909–1980
  Planning and analysis of observational studies.

  (Wiley series in probability and mathematical
statistics, ISSN 0271-6356. Applied probability and
statistics)
  Includes bibliographies and index.
  1. Experimental design. 2. Analysis of variance.
I. Moses, Lincoln E. 1921–   . II. Mosteller, Frederick
1916–   . III. Title. IV. Series: Wiley series in
probability and mathematical statistics. Applied
probability and statistics section.

QA279.C63 1983   001.4′2   83-6461
ISBN 0-471-88719-6

Printed in the United States of America

10 9 8 7 6 5 4 3 2 1

Preface

In his notes on a *Monograph on Non-Experimental Studies*, W. G. Cochran wrote, before he had proceeded very far with the outline, "My plan has been for a short book (e.g., 150 pages) addressed not to statisticians but to subject-matter people who do or may do these studies. Because my experience lies there, the selection of topics and examples will be a bit oriented toward field studies in health, but I'll try to avoid too much of this. ... I'd like to keep it as simple as seems feasible. It will be more of a reference-type book than a text, but might be the basis of a seminar in a subject-matter department."

He spoke also of the difficulty of choosing an order for chapters in such a work. He thought the reader would have problems because, to prepare to read any chapter, some other chapters ought to be read. He felt this was symptomatic of the topic. The level of difficulty would vary a great deal from part to part, and the reader would find it necessary to adjust to this. He also found the problem that every author of a monograph has encountered: more research is needed in many spots "before I'll have something worth saying. I think, however, that if I pay too much attention to this point, it will never be written." Ultimately, he wrote six and a half of his intended seven chapters, and we present them here.

Cochran wrote his book on observational studies by assembling about a chapter a year, usually writing in the summer. He taught a course on the subject at Harvard using his notes. As the manuscript neared completion, his health suffered a sequence of blows, each of which required substantial time for recovery. His other extensive writings, both new research articles and revisions of Snedecor and Cochran's *Statistical Methods* and of his own *Sampling Techniques*, as well as his teaching, continued at a good pace in spite of these medical setbacks. Nevertheless, he did not get back to the book before his death, although he sometimes spoke of the possibility of getting someone to help complete it.

After his death, Cochran's wife, Betty, edited his collected papers. Mosteller consulted with her on the possibility that the manuscript on observational studies might be publishable. She searched through Cochran's papers and identified the manuscript. Nevertheless, the revision of the papers did not progress until Moses agreed to help with the editing. Since then it has been a joyous enterprise for both of us.

Several considerations encouraged us to undertake the work of editing the manuscript for publication. First, the planning and analysis of observational studies is an important area of statistical methodology. Second, Cochran made many strong contributions to this topic over his career, and he wrote from broad experience and with theoretical insight; surely his book should be published. Third, the manuscript itself was attractive, especially for its characteristic of considering failure of assumptions. Again and again Cochran gives attention to the behavior of a statistical procedure when one or more of the assumptions underlying its mathematical justification is false in some degree. Thus, regressions may not be linear; they may not be parallel; matching may be inexact; variances may be inhomogeneous; and so forth. The manuscript contained, as do many others of his publications, tables indicating the quantitative results of assumptions that failed by various amounts. These analyses both increase understanding of the statistician's tools and facilitate practical planning of studies.

The editors of a posthumous work have an obligation to explain the nature and extent of their interventions. We agreed that we would leave the material as originally written, except when there was clear need to make modifications. Once this decision was reached, we had a fairly straightforward path. Cochran did not write and repeatedly revise as some authors do, and so we felt justified in treating the manuscript as finished prose for the most part.

Cochran had told Mosteller repeatedly that the book was complete except for one chapter, but that the order of the chapters was still a puzzle. We had two candidates for the opening chapter—the slightly technical chapter that we have put first and the more chatty, unfinished Chapter 7. Some readers may wish to read Chapter 7 first. We chose to put it last because opening the book with a fragmentary chapter might give the reader the mistaken impression that much of the work was unfinished. Nevertheless, a completed Chapter 7 probably would have made a beautiful opening.

One topic was treated twice, both in Chapter 1 and Chapter 6. We removed most of this topic from Chapter 1 and placed some of it as the Appendix to Section 6.12. This move had the advantage of substantially lowering the technical level of Chapter 1.

We found occasional errors in formulas and corrected them. We also struggled to unify the notation, whose variety may have stemmed from chapters being written at disparate times and places. We are sure that Cochran would have corrected this in his own final review of the book.

We have preserved the economy of the technical writing; some brevity was achieved by relying on the reader to provide parallelism, by using implicit definitions, by saying in words what might require several subscripts, or by adopting occasional tricky notation that the reader will need to detect. Using these devices, Cochran avoids many ugly equations and mathematical expressions. We have decided that terms should usually be defined, and so we have added some definitions of expressions where the reader would otherwise have to guess at their meanings.

The references needed attention—their state was mixed from chapter to chapter. They were sometimes complete, or nearly complete, and sometimes were only indicated in the text by author or by author and a dubious date. We have not added new references, except possibly where we may not have understood the source intended. If we have not found the appropriate ones, we apologize and will appreciate having misjudgments brought to our attention. We have included with each reference to Cochran's own work, the number (in square brackets) which refers to the presentation of the referenced work in William G. Cochran, *Contribution to Statistics*, John Wiley & Sons, New York, New York, 1982. For some readers, this will be a more convenient source than the original publication.

Although we had the original outline for Chapter 7, Cochran did not keep to it and so we cannot conjecture what the rest of the chapter would have been like. Because Cochran expresses many personal views based on vast experience, it seemed wise to stop with his words.

LINCOLN E. MOSES
FREDERICK MOSTELLER

Stanford, California
Boston, Massachusetts
May 1983

Acknowledgments

In preparing the manuscript, we have been aided by Nina Leech, Marjorie Olson, and Beverly Weintraub, who have corrected and typed repeated versions with great care. Cleo Youtz helped us enormously with the references, as did Elaine Ung. John Emerson, Katherine Godfrey, Katherine Taylor Halvorsen, David Hoaglin, Marjorie Olson, and Cleo Youtz also advised us about other matters. Donald B. Rubin has kindly given permission for the reproduction of parts of tables from some of his work on matching and adjustment.

William Cochran's original work was partly facilitated by a grant from the National Science Foundation, and the preparation of this manuscript has been partly facilitated by National Science Foundation Grant No. SES 8023644.

L. E. M.
F. M.

Contents

CHAPTER 1

Variation, Control, and Bias

1.1 INTRODUCTION

This book deals with a class of studies, primarily in human populations, that have two characteristics:

1. The objective is to study the causal effects of certain agents, procedures, treatments, or programs.
2. For one reason or another, the investigator cannot use controlled experimentation, that is, the investigator cannot impose on a subject, or withhold from the subject, a procedure or treatment whose effects he desires to discover, or cannot assign subjects at random to different procedures.

In recent years the number of such studies has multiplied in government, medicine, public health, education, social science, and operations research. Examples include studies of the effects of habit-forming drugs, of contraceptive devices, of welfare or educational programs, of immunization programs, of air pollution, and so forth. The growing area of program evaluation has also caused more studies. Everywhere, administrative bodies —central, regional, and local—devote resources to new programs intended in some way to benefit part or all of the population, or to combat social evils. A business organization may institute changes in its operations in the hope of improving the way the business is run. The idea has spread that it is wise to plan, from the beginning of the program, to allocate some of the resources to try to measure both the intended and any major unintended effects of the program. This evaluation assists in judging whether the

1

program should be expanded, continued at its current level, changed in some way, or discontinued.

Such studies will be called *observational*, because the investigator is restricted to taking selected observations or measurements that seem appropriate for the objectives, either by gathering new data or using those already collected by someone else.

The type of observational study described in this book is restricted in its objectives. The study concentrates on a small number of procedures, programs, or treatments, often only one, and takes one or more response measurements in order to estimate the effects. Examples include studies of the effect of wearing lap seat belts on the amount and type of injury sustained in automobile accidents; studies of the amount learned by people watching an educational television program; studies of the death rates and causes of death among people who smoke different amounts; and the National Halothane Study, which compared the death rates associated with the use of the five leading anesthetics in U.S. hospital operations. The objectives here are close to those in controlled experiments, leading some writers to call such studies either quasi experiments or the more disapproving term, pseudo experiments.

A basic difference between observational studies and controlled experiments is that the groups of people whom the investigator wishes to compare are already selected by some means not chosen by the investigator. They may, for example, be self-selected, as with smokers or wearers of seat belts; selected by other people, as in the choice of anesthetic used in an operation; or determined by various natural forces, as in the comparison of premature-birth children with normal-term children. In general, the investigator is limited to two choices. First, he may have a choice of different contrasting groups from which to draw his samples for comparison. For instance, for a comparison of people residing in heavily air-polluted areas with people living in relatively clear air, contrasting areas exist in different residential parts of many cities, and the investigator may select from several areas within the same city. Second, having selected contrasting groups, the investigator may have greater or less flexibility in the kinds of samples which can be drawn and measured for statistical analysis.

This class of observational studies may be distinguished from another class, sometimes called *analytical surveys*, that have broader and more exploratory objectives. The investigator takes a sample survey of a population of interest and conducts statistical analyses of the relations between variables of interest to him. A famous example is Kinsey's study (1948) of the relations between the frequencies of certain types of sexual behavior and the age, sex, social level, religious affiliation, rural–urban background, and social mobility (up or down) of the person involved. The Coleman Report

(1966), based on a nationwide sample of schools, dealt with the question: to what extent do minority children in the Unites States (Blacks, Puerto Ricans, Mexican-Americans, American Indians, and Orientals) receive a poorer education in public schools than the majority whites? As part of the study, an extensive statistical analysis was made of the relation between school achievement and characteristics of the school (e.g., teachers, facilities), the child's aspirations and self-concept, and the home background. Much descriptive information was also obtained about the extent of racial segregation, the types of school facilities, and so forth.

Such analytical surveys vary in the extent to which the primary interest is in causal relations. The relations discovered in the statistical analyses often suggest possible causal hypotheses, later to be investigated more directly in the observational studies of the first class. Sometimes, causal hypotheses that appear plausible are adopted as a basis for action. The long-term Framingham Study [Dawber (1980)] took a sample of about 4500 middle-aged men, made numerous measurements on each man of variables that might be related to the development of heart disease, and followed the men for years to discover which men developed heart disease. Men who were obese, heavy cigarette smokers, and with high blood pressure were found to have the highest frequency of heart disease. Along with other data, these results were influential in leading to attempted control of these three variables as a standard preventive measure in medical practice.

1.2 STRATEGY IN CONTROLLED EXPERIMENTS—SAMPLED AND TARGET POPULATIONS

Although our concern is not with controlled experiments, it is worthwhile, for two reasons, to consider in succeeding sections the strategy that has been developed in controlled experimentation with variable material. First, the problems that face the experimenter are, in general, the same as those that face the investigator in an observational study. Second, the controlled experimenter has more power to study causal effects, and the techniques developed have been more fully worked out and described. Thus we may consider the two questions: What aspects of the approach in controlled experimentation can usefully be borrowed for observational studies? What are the most difficult problems?

The field of agriculture, in which the modern technique of experimentation with variable material was first developed, is convenient for illustration. Suppose that the objective is to compare the yield per acre of a new variety of a crop with a standard variety. If the new (N) and the standard (S) variety are each grown on a number of plots in the same field, the yield per plot will be found to vary from plot to plot. This variation immediately raises the problem: How far can we trust the mean difference $\bar{y}_N - \bar{y}_S$ over n plots of each variety as an estimate of the superiority of the new variety? The experimenter knows that if he increased n or decreased it, the quantity $\bar{y}_N - \bar{y}_S$ would change.

After some false starts, this problem was finally handled roughly as follows. At least conceptually, an experiment could be so large that the difference $\bar{y}_N - \bar{y}_S$ would finally assume a fixed value, say $\mu_N - \mu_S$. The observed $\bar{y}_N - \bar{y}_S$ from n repetitions is regarded not as an absolute quantity, but as an *estimate* of the value $\mu_N - \mu_S$ obtained for the population of repetitions of the trials. The theory of probability and a simple mathematical model were then invoked to prove that under certain assumptions the estimate $\bar{y}_N - \bar{y}_S$ is normally distributed about $\mu_N - \mu_S$ with standard error $\sqrt{2}\,\sigma/\sqrt{n}$, where σ^2 is the variance in yield from plot to plot. Student later removed the difficulty encountered when the investigator does not know σ, by showing that under the same assumptions, $(\bar{y}_N - \bar{y}_S)/(\sqrt{2}\,s/\sqrt{n})$ follows the t distribution, where s is the estimate of σ from the experiment. (The rules for calculating s and its number of degrees of freedom depend on the detailed structure of the experiment.) This theory led to tests of significance and confidence intervals for $\bar{y}_N - \bar{y}_S$ as tools in the interpretation of the results.

In the theory that led to these results, one basic assumption required is that the repetitions in the experiment are a random sample of the population of repetitions. To put it the other way round, statistical inferences about $\bar{y}_N - \bar{y}_S$ by these methods apply to the population of repetitions of which the experiment is a random sample. This immediately raises the question: Is this the population to which the experimenter would like the results to apply? The answer must usually be "no," particularly when the n repetitions completely fill the field, so that the population of indefinitely many repetitions is purely conceptual. In experiments and observational studies, the terms "the sampled population" (to denote this population of repetitions) and "the target population" (to denote the population for which the objective of the research is to make inferences) are useful. In experimental research the sampled population is nearly always much narrower and more restricted than the target population. Thus a medical experiment on the treatment of a disease may be done on the patients having the appropriate diagnosis who turn up in a certain ward or clinic of a certain hospital in a certain six-month period. Experiments in behavioral psychology are often conducted using graduate students, and other volunteer students (paid or unpaid) in a university's psychology department. The target populations may be all patients in a certain age range with this diagnosis or all young persons in a certain age range.

Experimenters seldom have the resources or the interest (this is not their area of expertise) to conduct their experiments on a random sample of the target population. A partial exception occurs in certain problems in agriculture. For instance, the initial comparisons of the new (N) and the standard (S) varieties may be done on small plots at an agricultural experiment station. Small plots are used because with good design σ becomes smaller and n larger on a given area of land, making $\sqrt{2}\,\sigma/\sqrt{n}$, the standard error (SE), smaller. Uniform fields and good husbandry also lead to decreased σ. It is known, however, that comparisons $\bar{y}_N - \bar{y}_S$ for the sampled population of small plots at an experiment station (which often achieves higher than average yields) may not necessarily hold for the target population of farmers' fields in which the farmer may consider replacing S by N. Consequently, the purpose of the small-plot experiment is sometimes regarded as primarily to pick out promising new varieties. An N which beats S convincingly at the experiment station will then be compared with S on more nearly life-size plots at a sample of farmers' fields. This sample is seldom drawn strictly at random, since both willingness and some skill are required of the farmer; but the objective is to sample the range of conditions that occur in farmers' fields. These trials may be continued for several years to sample climatic variations.

In this example, moving from the sampled to the target population involves very substantial additional expenditure of resources and time after the original experiment or experiments have finished. Sometimes a step in this direction is taken by cooperative work between investigators of the same general problem. Cooperative experiments on the treatment of leprosy, for instance, were conducted simultaneously, with the same plan, treatments, and measurement of response, at leprosaria in Japan, The Philippines, and South Africa, while the same has been done on rheumatic fever in experiments conducted in England and the United States. With human subjects, a casual sampling of a broader population can also occur if the results of an initial experiment by an investigator excite interest. Other experimenters in different places with different subjects repeat the experiment, perhaps with slightly different techniques, to see if they get similar results. After a lapse of time, a more broadly based summary of such experiments may permit conclusions more nearly applicable to the target population.

It is sometimes stated that observational studies are often in a stronger position than experiments, with regard to the gap between sampled and target population. Analytical surveys may collect for analysis a random sample of the actual target population so that, apart from problems of nonresponse, there is no gap. In the restricted causally oriented studies the situation varies. We may have to take a group of persons subject to some program and a comparison group not subject to the program where we can find them. But sometimes, particularly in studies made from records, we can start with random samples of the immediate target population in constructing treated and nontreated groups.

More will be said on this problem later. At a minimum, the investigator should be aware of the nature of the target population when he selects comparison groups and should describe as clearly as possible the nature of the population that he believes he has sampled.

In this problem the experimenter has another weapon—factorial experimentation—that can be used to some extent in observational studies, provided that the composition and sizes of the samples are appropriate. In thinking whether to recommend N or S to the farmer, the experimenter knows that some farmers will apply one, two, or three of the common fertilizers, say sulphate of ammonia (S.A.) (supplying nitrogen), super phosphate [supplying phosphorus (P)], or potassium chloride [supplying potassium (K)]. Some sow the seed at heavier rates than others. Should the recommendations depend on the individual farmer's practices?

In dealing with this problem the experimenter might use what is called a 2^5 factorial experiment. There are now $2^5 = 32$ treatments, consisting of all combinations that can be made from the two levels of each factor, perhaps

$$\left\{ \begin{array}{c} \text{No S.A.} \\ \text{S.A.} \end{array} \right\} \left\{ \begin{array}{c} \text{No P} \\ \text{P} \end{array} \right\} \left\{ \begin{array}{c} \text{No K} \\ \text{K} \end{array} \right\} \left\{ \begin{array}{c} S_1 \\ S_2 \end{array} \right\} \left\{ \begin{array}{c} \text{Variety } N \\ \text{Variety } S \end{array} \right\}$$

A single replication of the experiment now requires 32 plots instead of 2. But within this replication N and S have been compared separately in each of the 16 combinations of levels of the other four factors. Thus for the *average* difference between N and S we obtain 16 comparisons from the 32 plots, just as if the experiment was nonfactorial but had 16 replications. The same is true of the average effects of S.A., P, and K and for the coverage difference between the two seeding rates S_1 and S_2.

We come to the question: Is the difference $\bar{y}_N - \bar{y}_S$ affected by the presence or absence of S.A.? The experimenter has 8 comparisons of N and S when S.A. is not applied and 8 comparisons when S.A. is applied, and their averages are comparable with respect to P, K, and the seeding rates. The experimenter can therefore estimate and test for significance the difference

$$(\bar{y}_N - \bar{y}_S)_{\text{S.A.}} - (\bar{y}_N - \bar{y}_S)_{\text{No S.A.}}$$

The same type of comparison can be made with respect to the effects of P, K, and different seeding rates on $\bar{y}_N - \bar{y}_S$, though the sample sizes per single replication are now 8 instead of 16. This type of experiment and the resulting comparisons greatly help to broaden the basis for recommendations from experiments.

Consider an observational study in which a program is made available to certain subjects (P) but not to others (O). In trying to estimate the effect of the program on y, the investigator may wonder: Does the effect differ for men and women, for older and younger subjects, for persons with incomes above or below a certain level? In his statistical analysis he can compare $\bar{y}_P - \bar{y}_O$ in each of the eight subsamples formed by the combinations of the two levels of each of these other variables. The investigator then proceeds to estimate $\bar{y}_P - \bar{y}_O$ separately for men and women, older and younger persons, richer and poorer persons. Two difficulties arise: (1) The investigator will probably have unequal numbers of P and O subjects in each subsample, so that the analysis is more complex, involving multiple classifications with unequal numbers. (2) In some cells the numbers of P and O subjects may be so small that the comparisons of interest have poor precision and nothing very definite is learned. Nevertheless, it is useful to consider such variables as sex, age, and income, both in selecting the sample and in the analysis, and to try to determine the importance of attempting analyses of this kind, which may lead to sounder conclusions.

1.3   THE PRINCIPAL SOURCES OF VARIATION IN THE RESPONSES

To return to the discussion of strategy in experiments, the fact that the response y varies from plot to plot under both N and S implies, of course, that y is influenced by variables other than the treatment N or S. Commonly, there are numerous such sources of variation. Investigators, either in experiments or observational studies, who write down the sources that they know or suspect often end with an impressively long list. In considering such sources, the investigator may think of them as falling into one of three classes:

1.   Sources whose effects the investigator tries to remove, wholly or partly, from the comparison $\bar{y}_N - \bar{y}_S$ by control during the course of the experiment or in the statistical analysis of the results.

2.   Sources whose effects the investigator handles by randomization and replication. Randomization, unlike control, does not attempt to *remove* the effect of a source of variation, but instead makes this source act like a random variable, equally likely to favor N or S in any repetition. Consequently, if a given source of variation contributes an amount with standard

deviation, σ_1, to the variation in y, the contribution of this source to the SE of $\bar{y}_N - \bar{y}_S$ is $\sqrt{2}\,\sigma_1 / \sqrt{n}$. By randomization and replication the contribution of any source can be made small if n is large enough.

3. Sources whose effects are neither controlled nor randomized. In an ideal experiment there should be no sources of this kind, and experimenters often forget the possible existence of these sources. If such sources happen to act like randomized variables in class 2, their effect is merely to increase σ. If, however, they are related to (confounded with) the treatment difference, they may give misleading results for which tests of significance are no protection. Well-known examples occur in medicine. If both the patient and the doctor measuring the response y know which treatment the patient has received, this may produce a biased overestimate $\bar{y}_N - \bar{y}_S$, especially if N is a new drug with an impressive name and S is simply bed rest. Whenever possible, medical experimenters go to considerable trouble to conduct "double blind" experiments, in which neither the patient nor the doctor measuring the response knows the treatment being measured for the patient. Another instance is the "novelty" effect. N may do well in the first experiment because it is a change from the usual routine, but later experiments (when N and S are both familiar) may show little difference. Sometimes a bias is introduced because of a wrong decision by the experimenter. Suppose that the measurement of y requires a complex laboratory analysis on a sample of the subject's blood, and that n is large enough so that two laboratories must be used. The experimenter sends all samples from N to lab 1 and all samples from S to lab 2. Any consistent difference between labs in the levels of y found when analyzing the same sample (and such differences are not uncommon) contributes a bias to $\bar{y}_N - \bar{y}_S$.

[Kish (1959) gives an excellent discussion of these sources of variation as they affect experiments and observational studies.]

1.4 METHODS OF CONTROL

In considering the variables whose effects on y should be removed by control, one might at first advise "so far as your knowledge of y permits, select for control those x variables that are the major contributors to the variation of y." Thus if y has a linear regression in the sampled population on each relevant x variable, this advice leads to selecting for control the x whose squared correlation ρ^2 with y is highest. If the regression model for some variable x is

$$y = \alpha + \beta(x - \mu) + e$$

where e is the residual term, it follows that

$$\sigma_y^2 = \beta^2 \sigma_x^2 + \sigma_e^2 = \rho^2 \sigma_y^2 + \left(1 - \rho^2\right) \sigma_y^2$$

Successful removal of the effect of x by control reduces σ_y^2 to $\sigma_e^2 = (1 - \rho^2) \sigma_y^2$, the reduction being greatest when ρ^2 is greatest. But the decision to control or not to control with regard to x depends also on the inconvenience and expense that is required to control. With some variables it might be better in experiments to randomize with respect to x, thereby reducing its contribution to the SE of our comparison by extra replication.

The devices for control fall under three headings:

1.   *Refinements of Technique.*   Examples include use of intricate measuring instruments that reduce errors of measurement of y; of instruments that maintain constant temperature, humidity, and light throughout a laboratory; of experimental animals specially bred for uniformity (or of fields selected for test uniformity of yield). As has been noted, some of these refinements make the sampled population much more restricted than the target population. In general, such devices merely attempt to follow an old maxim for precise experimentation: "Keep everything constant except the difference in treatments." With variable material this maxim cannot be followed completely and the experimenter must decide to what extent he will try to follow it.

In observational studies there are some opportunities of this type also, particularly with standards of measurement and the choice between less- or more-variable groups in which to conduct the study. For instance, in the *Midtown Manhattan Study* (1962, 1975, 1977) interviews among male workers in New York City were used to conduct an analytical survey of the relationship between undiagnosed mental illness and various characteristics of the worker and his background. The investigators planned to use trained psychiatrists for the difficult measurement problems, but found that the scarcity of such specialists would limit their sample to a size much too small for their planned statistical analyses. Kinsey (1948) faced a similar conflict. Although his planned male sample size was 100,000, his high standards for the selection and training of interviewers restricted the interviewing force to a very small number.

2.   *Blocking and Matching.*   The idea is to arrange the experiment in separate individual replications and to try to keep the variables to be controlled constant *within each replication*. In this way, precise comparisons among the treatments can be made even if the controlled variables are far from constant throughout the available samples. In agricultural field experiments the replication is usually a compact block of land—hence the name

blocking—since it had been found that small plots close together tend to give similar yields. Additionally, operations like plowing, harvesting, and weighing the plot yields are applied in the same way at the same time within a block. Randomization is used in allotting the treatments to the individual plots in the block, with independent randomizations in each block.

The same blocking can be used to control several sources of variation simultaneously. For instance, an experiment was conducted to investigate whether injection of a certain chemical into rats would enable them to better withstand exposure to a poison gas. The two members of the same block were litter mates of the same sex, which made them of the same age and similar genetic constitution. They were put into a bell jar together into which poison gas was fed, the measurement being time to death, so that variations from trial to trial in the rate of feeding the gas affected each treatment equally. The two rats varied a little in weight, but random allocation of the treatment to the rats within a block made any advantage from selecting the heavier rat average out.

This technique is widely employed in observational studies, usually under the name *matching*. In comparing two groups of people under different programs, the investigator may judge that the response y will be affected by the age, sex, and educational level of the subjects. In matching, the investigator tries to find pairs of the same or nearly the same age, the same sex, and similar educational level. Matching may be performed with respect to a single variable or occasionally to as many as a dozen variables.

3. *Control During the Statistical Analysis.* This method is also widely used in observational studies. If the variables to be controlled are all classifications like sex, Republican, Democrat, or Independent, this control usually involves first calculating $\bar{y}_N - \bar{y}_S$ in each cell formed by these classifications as mentioned previously, and if appropriate, taking some weighted mean of the $\bar{y}_N - \bar{y}_S$ values. Since the controlled variables are constant within each cell, they do not affect this weighted mean.

If y and the x variables to be controlled are continuous, the investigator first constructs a regression model describing how the mean value of y depends on the x's. This method is often called the *analysis of covariance*. In the simplest case of a single x and a linear regression, the model for the sampled population in an experiment is (before any treatment effect is added)

$$y = \mu_y + \beta(x - \mu_x) + e$$

where e is a residual with mean 0 for any fixed x, representing the combined effects of uncontrolled randomized variables. If the treatments have effects

τ_1 and τ_2, it follows that

$$\bar{y}_1 - \bar{y}_2 = \tau_1 - \tau_2 + \beta(\bar{x}_1 - \bar{x}_2) + \bar{e}_1 - \bar{e}_2$$

The error in $\bar{y}_1 - \bar{y}_2$ as an estimate of $\tau_1 - \tau_2$ has a part $\beta(\bar{x}_1 - \bar{x}_2)$ due to this x variable and a part $\bar{e}_1 - \bar{e}_2$ due to other variables. Removal of the effect of x is done by computing a sample estimate b of β by standard regression methods, and replacing $\bar{y}_1 - \bar{y}_2$ by the adjusted estimate

$$(\bar{y}_1 - \bar{y}_2) - b(\bar{x}_1 - \bar{x}_2)$$

Since b will be subject to a sampling error and will not exactly equal β, the removal is not quite complete, but in large samples will be nearly so when all other assumptions hold. By use of multiple regression the method can adjust ·for more than one x variable and for curvilinear relations by including terms in x^2. In both experiments and surveys, regression adjustments for some variables can be combined with blocking or matching for others.

To summarize Sections 1.3 and 1.4, the experimenter tends to think of three types of variable, other than treatment differences deliberately introduced, that may affect the response y: (1) variables whose effects the experimenter tries to remove from $\bar{y}_1 - \bar{y}_2$ by control devices like blocking or adjustments in the analysis; (2) variables whose effects will be averaged out by randomization and replication, so that they create no systematic error or bias in $\bar{y}_1 - \bar{y}_2$ and that their effects are taken into account in the standard error associated with $\bar{y}_1 - \bar{y}_2$; and (3) variables neither controlled nor randomized. In some cases the experimenter believes that these variables act like randomized variables. Thus in some industrial experiments it is convenient and cost-effective to conduct and measure all replicates of a given treatment consecutively rather than randomizing this order among treatments. From the experimenter's knowledge of the chemical processes involved, the experimenter may argue that he sees no reason why this failure to randomize produces any systematic error in the comparison of the treatment means. In experiments in which numerous physical operations are required, the experimenter may argue that randomization at certain stages is a needless and perhaps troublesome step, although sometimes a single randomization, if planned from the beginning, can handle many variables simultaneously. In other cases devices like blindness may remove a bias likely to be related to the particular treatment, thus placing the variable in class (1) instead of (3).

In observational studies, randomization can sometimes be introduced at certain stages, with or without blocking. If several judges have to be employed to make a difficult subjective rating from the subjects' question-naires, each judge might be assigned an equal-sized subsample of subjects from each treatment group, rated in random order to protect against systematic changes in the judge's levels of ratings, as has been claimed to occur in the marking of examinations. But such randomization usually handles only a few of the variables that affect y. Thus in observational studies there are normally only two classes—variables for which control is attempted and variables neither controlled nor randomized.

The techniques for control—matching and adjustment—therefore assume a more prominent role in observational studies than in experiments. Chapters 5 and 6 treat these topics systematically.

1.5 EFFECTS OF BIAS

For both theoretical and practical reasons, presented more fully in later chapters, the investigator may do well to adopt the attitude that, in general, estimates of the effect of a treatment or program from observational studies are likely to be biased. The number of variables affecting y on which control can be attempted is limited, and the controls may be only partially effective on these variables. One consequence of this vulnerability to bias is that the results of observational studies are open to dispute. Such disputes, often voluminous, may contribute little to understanding. One critic may believe that failure to adjust for x_4 made the results useless, while the investigator may believe that there is little risk of bias from x_4.

The investigator must use his judgment, assisted by any collateral evidence that he can find, in appraising the amount of bias that may be due to an x variable. This judgment is needed both in planning the variables to be controlled and in drawing conclusions. Often, the direction of a bias can be predicted. A television station may give a test to volunteer subscribers after an educational program on a certain topic, to estimate the amount learned from viewing the program. The volunteers included (1) some who viewed the program and (2) some who did not. It can usually be assumed that those who elected to see the program already knew more about the topic prior to the broadcast than those who did not see the program; therefore $\mu_{1y} > \mu_{2y}$. In studies of the possible inheritance of some forms of cancer, cancer patients may be better informed about cancer among relatives in the preceding generation than noncancer subjects.

In drawing conclusions, the investigator can sometimes reach a fairly firm judgment that the maximum bias is small relative to $\hat{\tau}_1 - \hat{\tau}_2$. This may have been the situation in studies of the relation between frequent cigarette smoking and the death rate from lung cancer. Cigarette smokers are self-selected, and numerous possible sources of bias in comparing them with nonsmokers have been mentioned in the literature, with supporting data from samples of the two populations. But the increase in the lung-cancer death rate for frequent cigarette smokers is so large that it is difficult to account for more than a small part of this increase by other differences in the two samples.

1.6 SUMMARY

Comparative observational studies in human populations have two distinguishing features: they address causal effects of certain treatments, and the data come from subjects in groups that have already been constituted by some means other than the investigator's choice.

Characteristically, the population from which samples are taken (the sampled population) is narrower than the population for which conclusions about treatment comparisons are desired (the target population). This commends that the investigator should give an account of the population sampled and how it may differ from the target population.

Issues are somewhat clarified by considering the process of arriving at (practical) conclusions in agricultural research, even though that research is based on experimental, not observational, studies. Experiments at an agricultural research station comparing two varieties, say new (N) and standard (S), result in estimates of treatment differences and statistical significance that relate to the particular fields and weather at the research station. Generalization from that sampled population to the more varied fields, weather, and diverse farming practices of the countryside is advanced by additional studies, over years, on farms in that target population. But generalization can also be aided during the experiment at the research station by using experiments that employ various levels of important factors that vary among farmers, such as the use of different seeding rates and various fertilizers. Such experiments lead to comparing outcomes in subsets of the data defined by combinations of these deliberately introduced factors. Similarly, it can be useful in observational studies to make comparisons between the treatments in subgroups that correspond to various combinations of important variables such as age, income level, and sex.

In experiments, variables that contribute to variation in outcome can be dealt with by randomization or control (or else ignored). In observational studies, since control usually offers the only alternative to ignoring influential variables, methods of control are quite important. There are three general methods of control: (1) refinement of techniques through devices such as training interviewers and improving questionnaires, (2) blocking and matching, and (3) statistical adjustments, such as regression.

Control can usually be attempted on only a few of the many variables that influence outcome; that control is likely to be only partially effective on these variables. Thus the investigator may do well to suppose that, in general, estimates from an observational study are likely to be biased. It is therefore worthwhile to think hard about what biases are most likely, and to think seriously about their sources, directions, and even their plausible magnitudes.

REFERENCES

Coleman, J. S., E. Q. Campbell, C. J. Hobson, J. McPartland, A. M. Mood, F. D. Weinfeld, and R. L. York (1966). *Equality of Educational Opportunity* (2 vols.). Office of Education, U.S. Department of Health, Education, and Welfare, U.S. Government Printing Office, Washington, D.C., No. OE-38001, Superintendent of Documents Catalog No. FS 5.238:38001.

Dawber, T. R. (1980). *The Framingham Study: The Epidemiology of Atherosclerotic Disease.* Harvard University Press, Cambridge, Massachusetts.

Kinsey, A. E., W. B. Pomeroy, and C. E. Martin (1948). *Sexual Behavior in the Human Male.* Saunders, Philadelphia.

Kish, L. (1959). Some statistical problems of research design. *Am. Sociological Rev.*, **24**, 328–338.

Srole, L., T. S. Langner, S. T. Michael, M. K. Opler, and Thomas A. C. Rennie, with Foreword by Alexander H. Leighton (1962). *Mental Health in the Metropolis: The Midtown Manhattan Study.* Thomas A. C. Rennie Series in Social Psychiatry, Vol. I. McGraw-Hill, New York.

Srole, L. and A. Kassen Fischer, Eds. (1975). *Mental Health in the Metropolis: The Midtown Manhattan Study.* Book One, Revised and Enlarged. Harper Torchbooks, Harper & Row, New York.

Srole, L. and A. Kassen Fischer, Eds. (1977). *Mental Health in the Metropolis: The Midtown Manhattan Study.* Book Two, Revised and Enlarged. Harper Torchbooks, Harper & Row, New York.

CHAPTER 2

Statistical Introduction

2.1 DRAWING CONCLUSIONS FROM DATA

This chapter has two purposes. First, before considering difficulties in drawing inferences that are peculiar to observational studies, we will review some difficulties present to a greater or lesser degree, in all statistical studies, and will review standard techniques for handling them. This review is elementary and of the type given in introductory courses in statistics. Second, because of the limited amount of control that the investigator can exercise over data, a constant danger in observational studies is that estimates tend to be biased. Consequently, the last part of this chapter considers the effect of bias on a standard method of inference.

As a simple example, suppose that an investigator has a group of n persons exposed to some experience or causal force and a second group of n comparable persons who are not exposed to this force. After a suitable period of time, a measurement y of the response that is of interest is made on every person. Let y_{1j} and y_{2j} denote the values of y for the jth members of the first and second groups. As a measure of the effect of the exposure, the investigator takes $\bar{d} = \bar{y}_1. - \bar{y}_2.$, the difference between the means of the two groups. (The dot in the subscripts of $\bar{y}_i.$ indicates that we have averaged over the values of j.)

With human beings it is almost always found that even within a specific group, for example, the "exposed," the values of y_{1j} vary from person to person because of natural human variability. Consequently, $\bar{d}$ does not measure the effect of the exposure exactly, even in the simplest situation. Instead, $\bar{d}$ is more or less in error because of these fluctuations in the y_{1j} and y_{2j}. In handling this problem, the statistical approach starts by setting up a mathematical representation or model of the nature of the data.

First, we postulate that the n exposed people were drawn at random from a large population of exposed persons, and the n unexposed people from a large population of unexposed persons. Sometimes the data were actually obtained in this way. In comparing two types of workers A and B in a large factory, the investigator might have obtained a list of all the workers of types A and B in the factory. Then, with a table of random numbers, the investigator might have selected $n = 30$ persons of each type. Sometimes the data were obtained by random drawings from a small population that the investigator hopes is representative of a larger one. From a school system with 29 junior high schools, 16 might have been selected at random for the study, although the investigator's aim is to draw conclusions applicable to junior high schools across the nation. In many studies, however, there is no deliberate random drawing of the $2n$ people from larger populations. Some volunteering may be involved, or one of the two groups may be more or less a captive one. Some studies on hospital patients are based on those patients who are receiving treatment during the three weeks after the start of the study; the persons studied, for example, may be graduate students in psychology courses given by the investigator, or pupils in a school that agrees to cooperate with the investigator, or an unusual religious sect whose dietary habits interest a nutritionist.

Thus, in many studies the postulate is unrealistic that the data were drawn at random from specific populations. Nevertheless, the standard techniques of statistical inference, the best methods available at present for coping with this problem of person-to-person variability, apply only to the populations of which the data may be regarded as random samples. Consequently, the investigator who has a nonrandom sample must envision the kind of population from which the sample might be regarded as drawn at random. It is helpful to give a name—the *sampled* population—to this kind of population and to describe the ways in which it differs from the *target* population about which we would like to draw conclusions. From this we may be able to form some judgment about the ways in which these differences would alter the conclusions. These judgments are worth including in published reports, though they should be clearly labeled as such. This distinction between sampled and target populations will recur frequently throughout this book and is discussed further in Section 4.7.

To revert to the statistical analysis of the data on exposed and unexposed people, the simplest form of mathematical representation is as follows:

$$y_{1j} = \mu + \delta + e_{1j}; \qquad y_{2j} = \mu + e_{2j} \qquad (2.1.1)$$

In this model, μ is a fixed parameter representing the average level of response in the unexposed population. The parameter δ stands for the average effect of the exposure, therefore the mean of the first population is $\mu + \delta$. The quantities e_{1j} and e_{2j} are called random variables; they vary from one member of each population to another, and allow for the observed fact that the y_{1j} and the y_{2j} vary. It is assumed that over the respective populations the average values of the e_{1j} and the e_{2j} are both zero.

When we write down any mathematical model to be used as a basis for the analysis of data, it is essential to reflect on any assumptions implied by the model about the nature of our data. Analysis is likely to be misleading if derived from a model which makes erroneous assumptions. The simple model (2.1.1) implies an assumption that is at best dubious for most observational studies in practice. If δ is zero, it follows from (2.1.1) that y_{1j} and y_{2j} are drawn from populations having the same mean μ. In the simplest types of controlled experiment, the investigator often takes a step that is designed to ensure that this assumption holds. In order to form two samples of size n, one exposed and one unexposed, he first draws a sample of size $2n$ from a *single* population. He then divides this into two groups of size n by a process of randomization, usually from a table of random numbers. Since y_{1j} and y_{2j} initially come from the same population and differ only as the result of the randomization, this should guarantee the stated assumption.

In observational studies, however, exposed and unexposed groups are rarely found in this way. Nearly always, these groups are formed by forces beyond the investigator's control. People decide themselves whether to wear seat belts or to smoke; in the case of children, their parents decide whether to send them to public or private schools. Thus for observational studies a more realistic model is

$$y_{1j} = \mu_1 + \delta + e_{1j}; \qquad y_{2j} = \mu_2 + e_{2j} \qquad (2.1.2)$$

with $E(e_{1j}) = E(e_{2j}) = 0$ as before (where the operator E represents the operation of taking the expected value), but μ_1 is not assumed to be equal to μ_2 since the investigator has been unable to take any step to ensure this equality.

The simplest estimate of δ is the difference between the means of the two samples; that is, $\bar{d} = \bar{y}_{1.} - \bar{y}_{2.}$. If (2.1.1) can be assumed, $\bar{d} = \delta + \bar{e}_{1.} - \bar{e}_{2.}$ and the mean value of $\bar{d}$ in repeated sampling is δ. But from (2.1.2),

$$\bar{d} = \delta + (\mu_1 - \mu_2) + \bar{e}_{1.} - \bar{e}_{2.}$$

and the expected value of $\bar{d}$ is $\delta + (\mu_1 - \mu_2)$ instead of δ. We say that the estimate $\bar{d}$ is subject to a bias of amount $\mu_1 - \mu_2$. This indicates the reason for an interest in biased estimates in observational studies. Some illustrations of such sources of bias will be given in Section 2.4.

Reverting to the "no bias" situation with $\mu_1 = \mu_2$, even here $\bar{d}$ does not give us the correct answer δ, but an estimate of δ that is subject to an error of amount $\bar{e}_1. - \bar{e}_2.$. The best-known aids for answering the question "What can we say about δ?" are two techniques called the "test of significance" and the "construction of confidence intervals." The backgrounds of both techniques will be reviewed briefly.

2.2   TESTS OF SIGNIFICANCE

The test of significance relates to the question: Is there convincing evidence that exposure to the possible causal force has any effect at all? The question is not at all specific about the actual value of δ; it merely tries to distinguish between the verdict $\delta = 0$ and the verdict $\delta \neq 0$. The test requires some additional assumptions about the data. Suppose for simplicity that in their populations the e_{1j} and the e_{2j} both have the same standard deviation σ, though a test can be made without this assumption. If the e_{ij} are assumed to be normally and independently distributed, then theory says that $\bar{d}$ is normally distributed with population mean δ and standard deviation $\sigma\sqrt{2/n}$. The value of σ is not known, but an estimate s can be made from the pooled within-group mean square, where

$$s^2 = \frac{\sum(y_{1j} - \bar{y}_1.)^2 + \sum(y_{2j} - \bar{y}_2.)^2}{2(n-1)} \tag{2.2.1}$$

Furthermore, the quantity

$$\frac{\bar{d} - \delta}{s\sqrt{2/n}} \tag{2.2.2}$$

follows Student's t distribution with $2(n-1)$ degrees of freedom. Moderate departures from normality in the data have little effect on this result.

If δ were zero, we would calculate from the data the quantity $t' = \bar{d}/s\sqrt{2/n}$, and find that (2.2.2) and the statement following it show that t' would follow the t distribution. From the tables of the t distribution for $2(n-1)$ degrees of freedom, we calculate the probability P that a value of t as large as or larger than our calculated value t' would be obtained. If P is small enough, we argue that if δ were zero it is very unlikely that we would get a value of t as large as we did. We conclude that δ is not zero; that is, there was *some* effect of exposure. In practice, the most common dividing line between "small" and "not small" values of P is taken as $P = 0.05$, although no strong logical reason lies behind this choice. There is something to be said for reporting the actual value of P, particularly for the reader who wishes to summarize this investigation along with others, or wishes to form his own judgment as to whether δ differs from zero.

If P is not small, we have learned that a value of t as large as the observed value would quite frequently turn up if δ were zero. Such a result fails to provide convincing evidence in favor of the argument that δ differs from zero. Equally, the result by no means proves that δ *is* zero. Faced with the question "Is δ different from zero?" this result sits on the fence.

In practice, investigators use tests of significance in different ways. In some fields, the finding of a significant $\bar{d}$ has been regarded as necessary evidence which an investigator must produce to verify that an agent has an effect. This use has probably been beneficial in research. Some investigators with a wide fund of ideas have a fondness for making claims based on little solid work. The significance-level "yardstick" encourages them to produce firmer evidence if they want their claims to be recognized.

The finding of a nonsignificant $\bar{d}$, on the other hand, is often regarded as proof that $\delta = 0$. This conclusion has no logical foundation; it may have been suggested by the jargon "we accept the null hypothesis" often used in statistical teaching. If faced with a nonsignificant $\bar{d}$ in a test of an agent A, the investigator may decide to act as if the effect δ of A is zero or small, so that he drops any study of A and proceeds to some other agent that looks more promising. This decision, however, should be based on the investigator's judgment about the alternatives available. In fact, the probability that $\bar{d}$ is nonsignificant depends primarily on the smallness of the quantity $\delta\sqrt{n}/\sigma$, where n is the size of each sample and σ the standard deviation within each population. If we want to form a judgment about δ in the light of a nonsignificant $\bar{d}$, the values of n and σ are both relevant. Fortunately, this judgment is aided by the construction of confidence limits for δ, as discussed in Section 2.3.

There are two forms of the test of significance: the two-tailed and the one-tailed forms. In the *two-tailed* form, used in practice more frequently, we calculate the absolute value of t' (denoted $|t'|$), ignoring its sign. In the t table we look up the probability of getting a value of t greater than the observed t' in either direction; that is, in mathematical terms, the probability that $|t|$ exceeds the observed $|t'|$. (For the vertical bars read: "absolute value of".) The two-tailed test is appropriate when our initial judgment is that the true effect δ could be either positive or negative. In this event a value of δ far removed from zero can reveal itself either by making t' large and negative or by making t' large and positive.

The *one-tailed* test is appropriate* only when we know in advance what sign δ must have if it does not equal zero. In a teaching program designed to increase a child's knowledge of a certain subject, application of a one-tailed test implies that the program either increases the child's knowledge or has no effect; it cannot possibly decrease knowledge. When $\bar{d}$ is in the anticipated direction, the values of P in a one-tailed test are exactly half those in a two-tailed test having the same t'; therefore for given $\bar{d}$ a verdict that δ is unlikely to be zero can be reached for smaller sample sizes. When $\bar{d}$ is in the wrong direction, the conscientious user of a one-tailed test does not compute t'; rather the user concludes that the result does not justify rejection of the idea that δ is zero.

Some investigators misuse one-tailed tests. Before beginning the study, they are convinced that δ must be positive. If $\bar{d}$ is positive, they apply a one-tailed test as planned; if $\bar{d}$ is negative they apply a two-tailed test, having now recognized, perhaps reluctantly, that δ could be negative. The net effect is to make the actual significance probability level 1.5 times the announced significance level; for example, if the test is announced as being at the 5% level, it is actually at the 7.5% level.

2.3 CONFIDENCE INTERVALS

Confidence intervals relate to the question "How large is δ?" We have an estimate $\bar{d}$, but recognize that this will be more or less in error. Once again the t distribution is used. Let $t_{0.025}$ be the two-tailed 5% level of t for $2(n-1)$ degrees of freedom (d.f.). We know that $t = (\bar{d} - \delta)/s\sqrt{2/n}$ follows the t distribution when model (2.1.1) holds. Hence, unless our samples are of an unusual type that turns up only once in 20 times,

$$-t_{0.025} \leqslant \frac{\bar{d} - \delta}{s\sqrt{2/n}} \leqslant +t_{0.025}$$

Rearrangement gives

$$\bar{d} - t_{0.025}s\sqrt{2/n} \leqslant \delta \leqslant \bar{d} + t_{0.025}s\sqrt{2/n} \qquad (2.2.3)$$

---

*The editors do not regard Cochran's interpretation of the one-tailed test as the only appropriate one. In testing composite hypotheses, the null hypothesis might include both zero and losses, with the alternative of interest including all possible gains. We may be looking for gains from innovations and not be much interested in following up losses. In such circumstances, we think a one-tailed test is appropriate.

Thus, apart from an unlucky 1 in 20 chance that made our samples unusually dissimilar to the sampled populations, δ lies somewhere between the two limits in (2.2.3), called the 95% confidence limits. The width of the interval between the lower and the upper limits, $2\sqrt{2}\, t_{0.025} s/\sqrt{n}$, is a random quantity; its distribution depends on the variability in the populations, on the size of the samples, and on the confidence probability. The interval for 80% confidence probability is about $\frac{2}{3}$ as wide as the 95% interval and that for 50% probability is about $\frac{1}{3}$ as wide.

To summarize, our state of knowledge about the size of δ may be expressed by an estimate $\bar{d}$ and a pair of confidence limits within which δ is likely to lie, with an attached confidence probability indicating how likely. This probability is verifiable experimentally by setting up normal populations whose means differ by a known δ, drawing repeated pairs of samples, and computing the limits for a specified confidence probability, say 80%, at each draw. The values of the limits will vary from draw to draw, but about 80% of them will be found to enclose δ.

In published reports of studies, confidence limits are seldom stated explicitly. A more common practice is to give $\bar{d}$ and its standard error, $s_{\bar{d}} = \sqrt{2}\, s/\sqrt{n}$. The reader may then calculate his own limits by use of a t table. The number of degrees of freedom in $s_{\bar{d}}$ should also be given, but if they exceed 30, the 50%, 80%, and 95% limits are approximately $\bar{d} \pm 0.65 s_{\bar{d}}$, $\bar{d} \pm 1.3 s_{\bar{d}}$, and $\bar{d} \pm 2 s_{\bar{d}}$, respectively.

These confidence limits also supply a two-tailed test of significance. If the 95% limits include 0, $\bar{d}$ is not significantly different from 0 at the 5% significance level. When a value of $\bar{d}$ is not significantly different from 0, it is worth examining the corresponding confidence limits. Sometimes, particularly with large samples, both limits are close to 0. For example, suppose that $\bar{d} = +0.3$, with 95% limits -0.6 and $+1.2$. In the context of the problem, it might be clear that even if δ is as large as 1.2, this is of minor practical importance. In this event the conclusion "exposure had no appreciable effect" would be justified as a practical approximation. On the other hand, with small samples and high variability, we might find that $\bar{d} = +0.3$ as before, but that the limits are -3.0 and $+3.6$. If values of δ as low as -3.0 and as high as $+3.6$ have important practical consequences of different kinds, a realistic conclusion is that the study did not succeed in delimiting the value of δ sufficiently to determine whether exposure has an important effect. In this event it would be risky to conclude that δ can be assumed to be 0.

To summarize, even under ideal conditions the information about the size of the effect supplied by a two-group study on human subjects is not as clear-cut as is desirable. But with the aid of techniques such as tests of significance and confidence intervals, and with careful thought about the implications of the results, we should be able to avoid serious mistakes in the conclusions. The logical ideas behind these techniques may be debated, but the techniques serve well if regarded as a guide to, and not as a substitute for, our thinking.

2.4   SYSTEMATIC DIFFERENCES BETWEEN THE POPULATIONS

When comparing samples from exposed and unexposed populations in an observational study, a frequent source of misleading conclusions, as mentioned in Section 2.1, is that the two populations differ systematically with respect to other characteristics or variables that affect the response variable y. This section illustrates a few of the many situations that occur. The first example comes from the large-sample studies of the relation between smoking and death rates of men (Cochran, 1968). In these studies, information about smoking habits (including type of smoking and amount smoked per day) was first obtained by a mail questionnaire to a large sample of men. The three types considered here are nonsmokers, smokers of cigarettes only, and smokers of cigars and/or pipes (the cigar and pipe smokers were combined because the sample numbers are smaller). After receipt of the questionnaires the investigators were notified of any deaths that occurred among the sample members in subsequent months. From the data supplied to the U.S. Surgeon General's Committee on Smoking and Health, Table 2.4.1 shows the death rates for the three groups of men in a Canadian, a British, and a U.S. study.

To take the figures at face value, it looks as if cigar or pipe smoking results in a substantial increase in death rates, the differences from the nonsmokers being statistically significant in all three studies. For cigarette smokers, the British study shows an elevated death rate, significant at just about the 5% level, but the Canadian and U.S. studies show no elevations of this magnitude.

Table 2.4.1.   Death Rates per 1,000 Person-Years

Smoking Group	Canadian (6 years)	British (5 years)	United States (20 months)
Nonsmokers	20.2	11.3	13.5
Cigarettes only	20.5	14.1	13.5
Cigars and/or pipes	35.5	20.7	17.4

We remember, however, that the groups being compared are self-selected. Before rushing out to warn the cigar and pipe smokers we should ask ourselves whether there are other characteristics affecting death rates or variables in which these groups might differ systematically. For men under 40, I think it is correct to say that there is no such variable known to have a *major* effect on death rates. For older men, age is a variable that becomes of predominating importance. The death rate rises a lot as age increases, with a steadily increasing steepness of slope. An obvious precaution is to examine the mean ages of the men in each group, as shown in Table 2.4.2.

In all three studies, the cigar or pipe smokers are older on the average than the men in the other groups. This is not surprising, since cigar and pipe smoking are more frequent among older men. In both the Canadian and U.S. studies, which showed death rates about the same for cigarette smokers and nonsmokers, the cigarette smokers are younger than the nonsmokers. Clearly, no conclusion should be made about the relation between smoking and death rates without taking steps to try to remove the effects of these systematic differences in ages among the groups. The investigators who directed these studies were well aware of this problem and planned from the beginning to handle it. The available procedures for dealing with disturbing variables of this type in the statistical analysis are discussed in Chapters 5 and 6.

The possibility of biases of this type is now widely recognized whenever groups are self-selected. Suppose a television station has a one-hour adult educational program and wishes to measure how much the viewers of this program have learned from it. The station maintains a representative list of viewers of its programs. After the program the station invites a random sample of viewers, some of whom have seen this program and some who have not, to take an examination intended to reveal what has been learned from the program. Even if all those invited conscientiously take the examination, the station recognizes that the computed $\bar{d}$ almost certainly overestimates the effect of the program, because those who chose to view this program were probably more interested in the subject and better informed about it from the beginning, and would do better in the examina-

Table 2.4.2.   Mean Ages (in Years) of Men in Each Group

Smoking Group	Canadian	British	United States
Nonsmokers	54.9	49.1	57.0
Cigarettes only	50.5	49.8	53.2
Cigars and/or pipes	65.9	55.7	59.7

tion even if they had not viewed this program. In fact, the station would regard the problem of circumventing this overestimation as the major obstacle in conducting the study.

In a second example the effect of self-selection is less clear-cut. Several studies have compared the health and well-being of families who move from slum housing into new public housing, with those of families who remain in slum housing. The objective is to measure the supposed beneficial effects of public housing. However, in order to become eligible for public housing, the parents of a family may have to possess both initiative and some determination in dealing with a bureaucracy. One might argue that these individuals may already possess a desire to rise in the world that might show up in better health and well-being when the response measurements are made. Insofar as such effects are present, they would increase $\bar{d}$ and the unwary investigator may attribute this to the beneficial effects of public housing.

This example illustrates another aspect of the problem. While admitting the argument in the preceding paragraph, an investigator might retort, "Why should this overestimation be sufficiently large to seriously distort the conclusions? Why should this initiative and determination have much effect on the number of colds Johnny catches next winter?" A critic might reply that such parents have high aspirations for their children and may take better preventive medical care of the children than do the slum parents in this study. The point is that in many studies, sources of potential bias of this type are either unavoidable or overlooked until too late and that we can only guess about the size of bias that is created. In such cases our judgment about the direction and amount of bias, even if slender, is worthwhile when conclusions are being drawn. Incidentally, in a Baltimore public housing study, Wilner et al. (1955) neatly attempted to avoid the bias in question by noticing that the list of processed and eligible applicants for public housing was much greater than the available space. Consequently, both the "public housing" sample and the "slum housing" sample were drawn from this list. The initial "slum" sample was made larger than the "public housing" sample, because it was known that as housing became available some of the "slum" families would move into public housing during the course of the study.

Sometimes systematic differences between the exposed and unexposed groups are introduced in the process of measuring the responses, especially when these measurements involve an element of human judgment. A consulting statistician soon observes that some investigators become emotionally interested in the causal force they are studying and they want the force to show some effects. This is not said in deprecation. It is the business of the research worker to form pictures of what the world is like, and an imaginative interest in one's pictures is conducive to good research. Consequently, the investigator, when measuring the responses, may find that the levels of measurement may change unconsciously when moving from the unexposed to the exposed group.

This danger is widely recognized in medical studies of the progress of patients. I once watched an expert on leprosy for an hour while he examined a patient in meticulous detail in order to measure the patient's progress during the preceding two-months' treatment in an experiment on leprosy. The examination was complex, since bacteriological, neurological, and dermatological symptoms are all involved. At the end of the exam, the doctor dismissed the patient and said, "I rate this patient *Much Improved*"—this being the highest category of improvement that the scale allowed. At this point a blabbermouth at the back of the room, who had the code sheet, said "You'll be interested to know, doctor, that the patient was on placebo"—a placebo being an inert drug with no bactericidal effect on leprosy, intended only as a measure of comparison for the real treatments. The immediate retort of the doctor was "I would never have rated that patient 'Much Improved' if I had known he was on placebo. Call him back." There was silence for about 30 seconds; members of the planning team either stared at the expert or the blabbermouth in mixtures of sorrow and anger, or engaged in silent prayer. Finally, the expert said in a low voice, as if arguing with himself: "No. Let it stand."

For this reason a standard precaution in medical studies, as in this one, is that the doctor who is measuring the response variable should not be informed of the treatment the patient is receiving. In some studies no workable precaution may occur to the investigators. In studies on the possible inheritance of some form of cancer, cancer patients may be better informed about cancer in their relatives than are noncancer controls, and hence report more relatives with cancer. In a study of the inheritance of neuroticism it was proposed to enlist both neurotic and nonneurotic subjects and obtain information about the parental generation by questioning each group of subjects about neuroticism in their parents. Bradford Hill (1953) remarks: "What the adult neurotic thinks of his father may not always be the truth."

2.5   THE MODEL WHEN BIAS IS PRESENT

When systematic differences between the exposed and unexposed groups are present, the simplest model appears to be

$$y_{ij} = \mu_1 + \delta + e_{1j}; \qquad y_{2j} = \mu_2 + e_{2j}$$

giving

$$\bar{d} = \delta + (\mu_1 - \mu_2) + \bar{e}_1 - \bar{e}_2 = \delta + B + \bar{e}_1 - \bar{e}_2.$$

where $B = \mu_1 - \mu_2$ represents the amount of bias in the estimate $\bar{d}$. The change from the "no bias" situation is that $\bar{d}$ now estimates $\delta + B$. While this is obvious, it emphasizes a point that investigators or their critics sometimes overlook. Once the presence of bias is admitted, they sometimes take a morbid view of the situation, implying that nothing can be learned about δ from $\bar{d}$. Actually, since we can rarely be certain in observational studies that estimates are completely free from bias, a not unreasonable view is that *all* estimates are biased to some extent in observational studies. The problem is to keep the bias small enough so that we are not seriously led astray in our conclusions.

We shall examine the effect of bias on the probability that $\bar{d}$ lies within the interval $(\delta - L, \delta + L)$, in other words, that $\bar{d}$ is correct to within given limits $\pm L$. With no bias, $\bar{d}$ is assumed approximately normally distributed with mean δ and standard deviation $\sigma_{\bar{d}}$. Hence, the quantity $(\bar{d} - \delta)/\sigma_{\bar{d}}$ is a standard normal deviate, and we calculate the probability α that $\bar{d} - \delta$ lies within $\pm L$ by setting $z = L/\sigma_{\bar{d}}$ and locating in the normal tables the probability α that a normal deviate z lies within the limits $\pm z_{\alpha/2}$.

When the mean of $\bar{d}$ is $\delta + B$, on the other hand, the normal deviate is $(\bar{d} - \delta - B)/\sigma_{\bar{d}}$. This equals $(-L - B)/\sigma_{\bar{d}}$ when $\bar{d} - \delta = -L$, and $(L - B)/\sigma_{\bar{d}}$ when $\bar{d} - \delta = +L$. If $B = fL$, these limits become $-L(1 + f)/\sigma_{\bar{d}}$ and $+L(1 - f)/\sigma_{\bar{d}}$ or $-z_{\alpha/2}(1 + f)$ and $+z_{\alpha/2}(1 - f)$. For a given probability in the "no bias" case, $z_{\alpha/2}$ is known; for example, $z_{\alpha/2} = 1.96$ for $1 - \alpha = 0.95$. Thus, given f and the "no bias" probability, we can read from the normal tables the corresponding probability when bias is present.

Table 2.5.1 shows the probabilities that $\bar{d}$ lies within $(\delta - L, \delta + L)$ for $P = 0.99, 0.95, 0.90, 0.80, 0.70$, and 0.60 and $f = B/L$ running from 0.1 to 1.0, by intervals of 0.1, and also $f = 1.5$ and 2.0. If f is less than 0.2, the reduction in the "no bias" probability is trivial. Even for $f = 0.5$, the reduction remains moderate, for example, from 0.95 to 0.84 and from 0.80 to 0.71. When $f = 1.0$, however, all the probabilities are reduced to 0.5 or less, and decrease steadily for higher f.

To put it another way, the effect of a bias of amount B cannot make the probability that $\bar{d}$ is correct to within $\pm B$ more than $\frac{1}{2}$, no matter how large the sample is. However, the probability that $\bar{d}$ is correct to $\pm 2B$ is decreased only moderately by the bias; the probability that $\bar{d}$ is correct to $\pm 5B$ is decreased only trivially. Some investigators prefer to express quantities like B and L as percentages of δ. In these terms, a 15% bias makes the probability that $\bar{d}$ is correct to within 15% of δ at most $\frac{1}{2}$, but reduces only moderately the probability that $\bar{d}$ is correct to within $\pm 30\%$.

For a given value of f, we might expect that a higher "no bias" probability will result in a correspondingly higher probability when bias is present. Table 2.5.1 shows this happens when $f \leqslant 1$, but for $f > 1$ the probabilities go in the opposite direction when bias is present. With $f = 1.5$, for instance, a "no bias" P of 0.99 is reduced to $P = 0.10$, but a "no bias" P of 0.60 is reduced only to 0.32. The explanation is that $\bar{d}$ in the biased case is an unbiased estimate of $\delta + B$, which lies outside the limits $\delta \pm L$. As the random-sampling error $\sigma_{\bar{d}}$ decreases in this case, we are doing a better job of estimating the wrong quantity $\delta + B$, but a poorer job of estimating δ. With $f = 2$, $\bar{d}$ is an unbiased estimate of $\delta + 2L$, and can fall in the desired interval $\delta \pm L$ only by making a negative error of more than L in estimating $\delta + 2L$. Thus with $f = 2$, the probability that $\bar{d}$ lies in the desired interval is $(1 - P)/2$, which for $P = 0.95$ gives 0.025.

The lessons from this example are important. Even without bias, other sources of variability impose a limit on the accuracy that can be attained with high probability in observational studies. These probabilities are not drastically reduced by bias, provided that B is substantially less than the limit of error L that can be tolerated. If an investigator takes pains to remove the effects of suspected sources of bias, the effect of undetected bias may be to reduce a presumed 95% probability of lying within prescribed limits to something like 60 or 70%. In such cases the deleterious effect of

Table 2.5.1.   Effect of a Bias of Amount $B = fL$ on the Probability P that $\bar{d}$ Lies within Limits $(\delta - L, \delta + L)$

No Bias	Probability (P)					
	0.99	0.95	0.90	0.80	0.70	0.60
$f = 0.1$	0.99	0.95	0.90	0.80	0.70	0.60
0.2	0.98	0.93	0.88	0.78	0.69	0.59
0.3	0.96	0.91	0.86	0.77	0.68	0.58
0.4	0.94	0.88	0.83	0.74	0.66	0.57
0.5	0.90	0.84	0.79	0.71	0.64	0.56
0.6	0.85	0.78	0.74	0.68	0.61	0.54
0.7	0.78	0.72	0.69	0.64	0.58	0.52
0.8	0.70	0.65	0.63	0.59	0.55	0.50
0.9	0.60	0.58	0.56	0.54	0.52	0.48
1.0	0.50	0.50	0.50	0.49	0.48	0.45
1.5	0.10	0.16	0.21	0.26	0.30	0.32
2.0	0.005	0.025	0.05	0.10	0.15	0.20

bias is not that it makes the results completely wrong, but that we do not know how far the results can be trusted. The most misleading situation is that of a relatively large bias when the samples are large. In this case $\bar{d}$ is likely to have a small standard error, so that the 95% confidence interval is narrow and we congratulate ourselves on our accurate results. The actual probability that δ lies within these limits may, however, be tiny, as the results for $f = 2$ indicate.

This example is also relevant when we come later to study the techniques for removing *suspected* bias in observational studies. There is evidence that in many circumstances the available methods are not fully effective. They remove some, hopefully most, of the bias, but leave a residual part. Our hope is that for this residual part, B/σ is substantially less than the value of L which makes our results useful.

On occasion it helps to think in terms of B/δ or of $B/\bar{d}$. Suppose we have been unable to remove or reduce a specific source of bias, and are making speculative calculations as to how large a bias from this source can be. We can sometimes reach a firm judgment that even under the most unfavorable circumstances, B/δ is bound to be small. This seems to be the situation in prospective studies of the relation between heavy cigarette smoking and the death rate from lung cancer. Cigarette smokers are self-selected, and numerous possible sources of bias in comparing them with nonsmokers have been mentioned in the literature. But the increase in the lung-cancer death rate for heavy cigarette smokers versus nonsmokers is so large that estimates of the bias, often admittedly speculative, all seem to result in relatively small changes in the estimated δ.

The preceding discussion has dealt with the effect of bias on attempts to estimate δ correct to within limits $\pm L$. It is also worth considering the effect of bias on a test of significance of the null hypothesis $\delta = 0$ in relation to sample size. If we can assume, that apart from the bias, we have independent random samples from the two populations, then $\sigma_{\bar{d}} = \sqrt{2}\,\sigma/\sqrt{n}$. Consider first a two-tailed test. The type-I error is the probability that $\bar{d}$ lies outside the limits $(-\sqrt{2}\,\sigma z_{\alpha/2}/\sqrt{n}, +\sqrt{2}\,\sigma z_{\alpha/2}/\sqrt{n})$. With a bias of amount B, this is easily found to be the probability that a normal deviate z lies outside the limits $(-z_{\alpha/2} - B\sqrt{n}/\sqrt{2}\,\sigma, z_{\alpha/2} - B\sqrt{n}/\sqrt{2}\,\sigma)$. Suppose that $B\sqrt{n}/\sqrt{2}\,\sigma = \pm 0.5$ in a 5% test, for which $z_{\alpha/2} = 1.96$. The limits are $(-2.46, 1.46)$, or $(-1.46, 2.46)$, and the probability of type-I error P is 0.079. With $B\sqrt{n}/\sqrt{2}\,\sigma = \pm 1$, $P = 0.170$, and with $B\sqrt{n}/\sqrt{2}\,\sigma = \pm 2$, $P = 0.516$. Regardless of the sign of the bias, the type-I error of a two-tailed test is increased by bias. If n is large, the type-I error is so distorted that the test becomes meaningless.

While I can give only an unsubstantiated opinion, things may not be this bad in practice. Careful precautions against bias might reduce B/σ to say

0.05. For two samples of sizes 50, 100, and 200, the P's for the type-I errors become 0.057, 0.064, and 0.079, respectively; only for samples larger than 200 does the distortion become intolerable.

The effect of bias on a one-sided test depends of course on the direction of the bias in relation to the direction of the one-sided test. If these directions are the same, that is, if the alternative hypothesis specifies $\delta > 0$ and if $B > 0$, bias increases the type-I error still more rapidly than it does for two-tailed tests. Consider a one-tailed 5% test that assumes $\delta \geqslant 0$, with $z_\alpha = 1.645$. For $B\sqrt{n} / \sqrt{2}\,\sigma = 0.5$, 1, and 2 (the figures given in the two-tailed example), the P's for the type-I errors become 0.126, 0.260, and 0.639, respectively. On the other hand, if B is negative in this situation, the type-I error is decreased. In fact, if we are sure that $\delta \geqslant 0$ and if $B\sqrt{n} / \sqrt{2}\,\sigma$ is negative and sufficiently large, we might detect the presence of bias by noting that our $\bar{d}$ would be significant in the wrong direction, a logical contradiction of the notion that $\delta \geqslant 0$, unless negative bias is the explanation.

Admitting the inevitability of bias and the difficulty of securing random samples in observational studies, some writers argue that the test of significance is useless for guidance in such studies. In large-sample studies this can be so, only because any difference large enough to be of interest is certain to be significant by a standard test that ignores bias. One frequently finds no mention of tests in the discussion of the results of such studies. Tests can also be useless if our results are subject to biases that are completely unknown in size and direction. However, as Kish (1959) points out, such biases would render ineffective *any* attempt to draw conclusions from observational studies, not merely tests of significance.

The positive attitude toward this problem is to exercise precautions against bias in the planning and analysis of observational studies, in the hope that the remaining bias will not greatly disturb type-I errors or confidence probabilities. In the interpretation of tests of significance, whether the verdict is significant or nonsignificant, any judgment that we can form about the direction and size of the remaining bias is clearly relevant.

2.6 SUMMARY

This chapter discusses some of the difficulties in trying to draw sound conclusions from the results of a study. Even in the simplest type of study—a comparison of a group of people exposed to some causal force and a second group not so exposed—the study provides only an estimate of the average effect of the causal force that is subject to error. In the interpretation of this estimated difference, two major statistical aids are *tests of significance* and *confidence limits*.

The test of significance relates to the question of whether there is convincing evidence of some real effect. Confidence limits supply estimated upper and lower bounds to the size of the effect. The finding of a nonsignificant difference by no means proves that there was no real effect. It merely reports that a difference as large as that observed could have occurred with nonnegligible probability without the presence of a real effect different from zero. The interpretation of a nonsignificant difference is often helped by calculating confidence limits. With small sample sizes or a highly variable population, the confidence limits may show that the study failed to measure the size of the effect accurately enough for competent decisions. The soundest conclusion is that more work on the problem is needed.

Further difficulties are often present. Tests of significance and confidence limits apply only to the population of which the data are a random sample. This population—the sampled population—is often difficult to describe accurately and may differ in one or more respects from the target population to which the investigator wants his conclusions to apply. A useful part of the investigator's summary of results includes a statement of known differences between sampled and target populations and a judgment about the extent to which these differences may affect the conclusions as they apply to the target population.

In observational studies the exposed and nonexposed populations are usually created by forces beyond the control of the investigator, and may differ systematically in respects other than that of exposure. The consequence is that the sample mean difference $\bar{d}$ does not estimate the real effect of exposure, but $\delta + B$, where B is a bias term caused by these other systematic differences. The presence of bias decreases the probability that $\bar{d}$ as an estimate of δ is correct within given limits $\pm L$. This reduction in probability is moderate if B/L is less than 0.5, as, for example, $B/\delta = 10\%$ and limits of accuracy $L/\delta = \pm 20\%$ are sufficient for our purpose. This result is encouraging. In statistical studies it is hard to ensure that bias has been completely eliminated, particularly in observational studies, but measures taken to control bias in planning and analysis may reduce B/L to a value that is small enough for practical decisions.

Similarly, a bias in either direction increases the probability of a type-I error in a two-tailed test of significance, sometimes to an extent that makes the test meaningless in large samples. This fact emphasizes the importance of exercising precautions against bias in the planning and analysis of observational studies and of using any judgments about the direction and size of the remaining bias in the interpretation of the results of tests of significance.

REFERENCES

Cochran, W. G. (1968). The effectiveness of adjustment by subclassification in removing bias in observational studies. *Biometrics*, **24**, 295–313 [Collected Works #90].

Hill, A. B. (1953). Observation and experiment. *New Engl. J. Med.*, **248**, 995–1001.

Kish, L. (1959). Some statistical problems of research design. *Am. Sociological Rev.*, **24**, 328–338.

Wilner, D. M., R. P. Walkley, and S. W. Cook (1955). *Human Relations in Interracial Housing.* University of Minnesota Press, Minneapolis.

CHAPTER 3

Preliminary Aspects
of Planning

3.1 INTRODUCTION

This chapter discusses some of the decisions that are faced in setting up an
observational study designed to compare a limited number of groups of
people. Since observational studies vary greatly, not all the points considered here will be relevant in a specific study; also, the issues that are most
troublesome in some studies have been omitted in this chapter. The groups
of people to be compared have been subjected to different experiences or
agents, whose effects are the object of interest. We list below some examples
of this type of study.

Experience or Agent	Effects or Response Variables
Wearing lap seat belts	Severity and type of injury in auto accidents
Head injury to child at birth	Performance in school
Viewing educational television program	Amount learned on the relevant topics
Urban-renewal program	Improvements by private landlords
Rise in taxes	Consumer spending and saving
Distribution of contraceptive device	Acceptability, birth rate
Town fluoridation of water	Status of children's teeth
Permissive or authoritarian kindergarten	Amount of quarreling among children
Smoking of cigarettes	Mortality and illness from specific causes

In controlled experiments the term *treatments* is often used for the agents which the experimenter applies in order to measure the agents' effects. Where appropriate we will continue to use this term to denote the different experiences or agents in different groups of people. The term *responses* will denote the measurements taken to throw light on the presumed effects of the treatments.

3.2   THE STATEMENT OF OBJECTIVES

Before planning actually begins, it is helpful to construct and have readily available as clear a statement as can be made about the objectives of the study. At first, the statement may have to be expressed in rather general language. As planning proceeds, the statement becomes helpful when decisions must be made about the treatments, responses, and other aspects of the conduct of the study, since one has an opportunity to check which choices seem most likely to aid in achieving the objectives of the study. In fact, without such a statement it is easy in a complex study to make later decisions that are costly and not particularly relevant to the original objectives; or worse still, the decisions may make the objectives actually harder to attain.

Some investigators like a statement in the form of a list of questions that the study is intended to answer; other investigators prefer a list of hypotheses about the expected causal effects of the agents. An example of the latter form occurs in a study by Buck et al. (1968) of the effects of the chewing of coca leaves by residents of four Peruvian Indian villages. This study lists three hypotheses:

(1) Coca, by diminishing the sensation of hunger, has an unfavorable effect on the nutritional state of the habitual chewer. Malnutrition and conditions in which nutritional deficiencies are important disease determinants occur more frequently in coca chewers than among controls.

(2) Coca chewing leads to a state of relative indifference which can result in inferior personal hygiene.

(3) The work performance of coca chewers is lower than that of comparable non-chewers.

These three hypotheses are a good example of the general type of preliminary statement; the authors also explain that these particular effect areas were selected because of a previous report on the consequences of coca chewing by a Commission of the United Nations. The statement does not specify either the treatments or the response variables, but provides a background against which alternative choices can be judged. Regarding the treatments, typical questions that arise in this type of study are as follows: Will this be a two-group study of coca chewers versus non chewers, or can any attempt be made to measure the level and duration of chewing per person, so as to give more-detailed knowledge of the dose–response relationship? Is it feasible to try to measure some effects separately for different kinds of people, for example, older and younger persons, or male and female? The choice of response variables obviously presents difficulties in this study, both in the number of possible variables that might be considered and in the question of the accuracy with which each variable can be measured.

This statement specifies only the direction of the effect of coca chewing on each of the three listed characteristics of chewers. From this statement, one might expect the statistical analysis of the results to consist mainly of tests of significance of differences between chewers and nonchewers, as indeed it does in the study cited. In initial studies a statement of directions of effects may be as far as the investigator feels able to go. But, reverting to the previous discussion of estimates and tests of significance (Sections 2.2 and 2.3), it is always worth considering how far the main interest should lie instead in estimating the sizes of the effects and forming some judgment about their importance.

The retort might be "Aren't the two objectives, estimation and testing significance, essentially similar? With a continuous response variable, the test criterion for two groups is $t = \bar{d}/s_{\bar{d}}$. This involves both the estimated size of difference $\bar{d}$ and its estimated standard error $s_{\bar{d}}$." This is so, but when the emphasis was concentrated on tests of significance, I have seen studies in which only the value of t, or only the value of χ^2 in the comparison of two percentages, was quoted in the presentation of results. The interested reader had to dig out the value of $\bar{d}$ from earlier parts of the studies, or found that $\bar{d}$ could not be calculated at all. There are also large-sample studies in which $\bar{d}$ was significant at the 1% level, and left the impression that the effect was large because of confusion between a highly significant difference and a large effect. Examination of $\bar{d}$ and $s_{\bar{d}}$, however, left the impression that the size of the effect was modest—perhaps of no great practical importance. Also, thinking in terms of the estimation of the sizes of effects tends to alert one more to possible sources of bias that must be handled than does thinking in terms of tests of significance.

This difference in point of view may also influence the approach to the statistical analysis. Suppose that there are four groups—the treatment appearing at four different levels, $a_1, \ldots, a_4$. For a test of significance, the investigator may use the F test in a one-way analysis of variance, or the χ^2 test in a 2×4 table if the response is a $0 - 1$ variate. From the viewpoint of estimation, the objective becomes that of describing in its simplest terms the mathematical relationship, if any, between $\bar{y}_i$ or $\hat{p}_i$ and a_i. A simple approach would be to examine whether a straight-line relation appears to hold between $\bar{y}_i$ or $\hat{p}_i$ and a_i, or between simple transforms of these variables. This approach often produces a summary response curve that is more informative, and also provides a more powerful test of significance of the reality of the relationship.

As each decision is made in the later stages of planning, it is not a bad idea to record at the time the reason for the decision. This will be useful when the final report must be written, often a long time later, and may also be useful when facing related decisions. Moreover, in observational studies, the investigator sometimes knows so little about the merits of alternatives that when a decision is made he does not feel that the decision was well-informed; it may have been little better than tossing a coin. If this weak foundation for the decision is not recorded, the investigator is tempted at a much later date to invent a rationalization as to why this decision was a brilliant one.

When the statement of objectives has been constructed in as specific a form as possible, it leads to a number of questions about the actual conduct of the study: What locale is to be chosen? Where is the study to be done? (Often the locale has been tentatively selected before the investigator drafts his statement of objectives, but in exploratory studies on new possible causal agents the need for a study may be evident before any suitable locale has been found or even sought.) What aspects of the treatments or causal agents will it be necessary to measure? What measures of the responses will be taken? What groups of subjects are to be compared in order to appraise the effects of the treatments? What sample size is needed?

This chapter will consider questions relating to measurements of the treatments and the responses. Questions relating to comparisons and sample sizes will be considered in Chapter 4 and later chapters. Comments on the choice of a locale are best postponed until these other aspects of planning have been discussed, and are also given in Chapter 4.

3.3 THE TREATMENTS

In the simplest situations the treatment is a specific event, for example, wearing seat belts or viewing an educational television program, and it is only a matter, in this case, of recording the subjects that wore seat belts or saw the program and those that did not. Often, however, the treatment is known to vary from subject to subject in amount or level on some scale. Regarding measurement of this level, at least three situations exist. (1) It may not be feasible to attempt any detailed measurement; the resources may permit only a comparison of a treated and an untreated group with no possibility of distinguishing between different levels. (2) The level of the treatment may be roughly the same for persons within the same subgroup of the study sample, but varies from one subgroup to another. Examples might be studies of the relationships between noise levels in different factory workshops (subgroups) and productivity, hearing losses, or irritability in the workers, or the relationship between air pollution and bronchitis in house-wives living in different parts of a town. (3) It may be feasible to consider measurement of the treatment level separately for each person.

In situations (1) and (2), consider a two-group exploratory study (treated versus untreated) where the primary goal is to discover whether there appears to be an effect that might receive more careful investigation later. If the investigator is seeking a suitable treated group, or subgroup, it is advantageous to find one where the average level of treatment is high, so that a marked contrast is obtained. With two groups, the contrast is essentially the difference between the means. The power of a statistical test tells us the probability that the observed values will produce a statistically significant result.

Treatments often have approximately linear effects, either beneficial or deleterious, over the range of levels encountered in practice. Suppose that the true response to an amount τ of the treatment is $\beta\tau$. With treated and untreated groups each of size n, the value of $\delta/\sigma_{\bar{d}}$, the quantity that primarily determines the power of the t test, is $\sqrt{n}\,\beta\tau/\sqrt{2}\,\sigma_y$. Since n and τ enter into this formula in the forms $\sqrt{n}$ and τ, it follows that doubling τ (finding a treated group with a level twice as high) is equivalent to quadrupling n from the viewpoint of the power of the t test. If for a given level of τ the power of a two-sided t test at 5% significance level is 0.44, implying a less than 50–50 chance of revealing a true difference between treated and untreated groups, doubling τ increases this power to 0.91.

Of course, other factors also enter. For example, since people do not live in badly polluted air by choice, two areas offering the largest contrast in degree of pollution within a town may also show the largest difference in variables such as economic level and access to regular medical care. Thus areas with a large contrast in levels of treatment may also present a large contrast in potential sources of bias. But my judgment is that despite this difficulty a large contrast is a wise choice in exploratory studies.

What should the relative sizes of the two groups be? If the response y to an amount τ is roughly linear, $y = \alpha + \beta\tau$, and if the within-group variances σ_1^2 and σ_2^2 are roughly equal, experience in controlled experiments recommends two groups of equal size $N/2$, untreated and high-level. This plan minimizes the standard error of the estimate of β, the average change in y per unit increase in τ, for a given total number N of subjects. Incidentally, if there is evidence that σ_1^2 and σ_2^2 are substantially different in the two-group case, the optimum sample sizes are $n_1 = N\sigma_1/(\sigma_1 + \sigma_2)$ and $n_2 = N\sigma_2/(\sigma_1 + \sigma_2)$.

What about three subgroups—one untreated and the others having a high and an intermediate level. A third group, placed at one-half the high level, adds nothing to the estimate of β under these conditions. If the three groups are now of sizes $N/3$, the net effect of using three groups having equal variance instead of the two extremes is to increase the variance of the estimate of β by 50%. A third group whose level is somewhere between 30 and 70% of the high level adds little to the precision of the estimate of β. Even if we retained the end groups at size $N/2$ and added a third group of size $N/2$ whose level is somewhere between 30 and 70% of the high level, we would add little to the precision of the estimate of β.

The role of an intermediate level is to provide a test as to whether or not the line is straight. If the levels of the treatment are 0, ah, and h, the test is made by calculating the contrast $\bar{y}_2 - (1 - a)\bar{y}_1 - a\bar{y}_3$, which should be zero apart from sampling error if the line is straight. A t test for detecting a curve response is obtained by dividing this quantity by its estimated standard error.

The case for three levels rather than two is stronger in observational studies than in controlled experiments. An ever-present danger in observational studies is that some source of bias which affects the comparisons between groups has not been rendered unimportant by the plan and method of analysis. However, it is reassuring to find that $\bar{y}_2$ for the intermediate level lies between $\bar{y}_1$ and $\bar{y}_3$, though this result could still be produced by a bias which happened to operate in the expected direction of the treatment effect. If $\bar{y}_2$ was found to be less than $\bar{y}_1$, while $\bar{y}_3$ was greater, we would reexamine carefully both the suspected sources of bias and the arguments which led us to think that the response would be roughly linear. For instance, in seeking comparison groups for radiologists (as a group exposed to a certain amount of radiation), Seltser and Sartwell (1965) deliberately chose two other groups: one group was ophthalmologists and otolaryngologists, who should have practically zero exposure, and the other group was physicians, who use x rays to some extent and have an intermediate level of exposure.

3.4 MEASUREMENT OF TREATMENT LEVELS FOR INDIVIDUAL PERSONS AND THE EFFECTS OF GROUPING

When the level of the treatment varies from person to person, it may be possible to measure the amount received by a given person more or less accurately. In an urban-renewal program in which sums were offered to repair rented houses, the amount given to each house might be known accurately. In studies of the aftereffects of the atom bomb dropped on Hiroshima, the dose of radiation received by each person has been estimated roughly from a person's memory of his location and the amount of nearby shielding when the bomb fell. In the smoking-health studies, the amount smoked by an individual may vary at different times, and both this and the number of years of smoking vary from person to person. The best single measure of the level of smoking is not clear even with full and accurate records. In the studies on smoking, the level used was a broad interval of number of cigarettes smoked per day (e.g., less than 10, 10–19, 20–39, 40 +), reported at the time when the study questionnaire was sent out.

Given an estimate a_i of the level of the treatment received by the ith person, one possibility is to analyze the relation between the response y_i and a_i using regression methods. An alternative (as in the studies on smoking) is to divide the range of a into a few classes, recording only the class into which each person falls. With two groups the sample is divided into upper and lower classes; with three groups, into low, middle, and high classes; and so forth.

When this method is under consideration, natural questions arise, such as: How many classes should be formed? Should they be made equal or unequal in numbers of persons? Is this method much inferior to the use of regression methods? Some answers can be given when there is a linear regression of y_i on a_i and the quantity of interest is the slope β of the line (average increase in y per unit increase in x). We assume that a total of N subjects are available so that with c classes the average class has N/c subjects. With the regression method, the estimate of β is $\hat{\beta} = \Sigma(y_i - \bar{y})(a_i - \bar{a})/\Sigma(a_i - \bar{a})^2$. We call the variance of this estimate based on the distributed values of a, $V(\hat{\beta})$. If, instead, two groups are formed, the estimate $\hat{\beta}_2 = (\bar{y}_2 - \bar{y}_1)/(\bar{a}_2 - \bar{a}_1)$. With three or more classes the regression of the $\bar{y}_c$ (class means of y) on the $\bar{a}_c$ (class means of a) is calculated to obtain the estimate $\hat{\beta}_3$ for this method. We call the variance of $\hat{\beta}_g$ based on the g groups $V(\hat{\beta}_g)$.

The values of $V(\hat{\beta}_g)/V(\hat{\beta})$ for $g = 2, 3, \ldots$ naturally depend on the shape of the distribution of the levels a_i. When this distribution is unimodal and roughly symmetrical, results for the normal distribution can be quoted, since these results have been calculated by D. R. Cox (1957) for a similar purpose. In Table 3.4.1 the total number of persons, H, is constant and assumed large. For $g = 2$ to 5, the table shows the optimum class sizes, the corresponding values of $V(\hat{\beta}_g)/V(\hat{\beta})$, and the values of $V(\hat{\beta}_g)/V(\hat{\beta})$ for equal-sized classes. The table shows the increases in variance of the estimate when grouping is used. Thus grouping in two equal groups increases the variance by 57%. Another way of thinking about these numbers is that the efficiency of the grouped estimate compared with the ungrouped estimate is the reciprocal of the numbers given in the last two columns of the table. Thus for two classes, the relative efficiency of $\hat{\beta}_2$ compared with $\hat{\beta}$ is 100/1.57 or about 64%.

We conclude from Table 3.4.1 that (1) It pays to use more than two classes, although not much is gained by using more than four classes. The minimum variance of $\hat{\beta}_g$ with four classes is only 13% higher than the minimum variance attained by regression under these assumptions. (2) The class sizes that minimize $V(\hat{\beta}_g)$ are unequal; the central classes are substantially larger than the outside classes. (3) Nevertheless, the use of equal-sized classes (number of subjects) is only slightly less effective than the best set of classes; for example, 1.16 compared with 1.13 for $V(\hat{\beta}_g)/V(\hat{\beta})$ with four classes.

Investigations of several smooth nonnormal distributions of the a_i (Cochran, 1968) suggest that the results in Table 3.4.1 can also be used as a guide in such cases. However, there is a hint that with some nonnormal distributions, grouping into classes gives values of $V(\hat{\beta}_g)/V(\hat{\beta})$ that are slightly lower than with the normal distribution, thus losing less precision relative to regression methods than when the a_i are normal.

Thus far we have considered the loss in precision due to grouping into classes when the distribution of levels of a is smooth and unimodal. The loss

Table 3.4.1. Comparison of the Variance of $\hat{\beta}_g$ for g Optimally Grouped Classes by Level of Treatment into Classes with the Variance of $\hat{\beta}$ Based on the Normal Distribution for a Linear-Regression Situation

Number of Classes g	Optimum Class Sizes	$V(\hat{\beta}_g)/V(\hat{\beta})$	$V(\hat{\beta}_g)/V(\hat{\beta})$ for Equal-Sized Classes
2	$0.5N, 0.5N$	1.57	1.57
3	$0.27N, 0.46N, 0.27N$	1.23	1.26
4	$0.16N, 0.34N, 0.34N, 0.16N$	1.13	1.16
5	$0.11N, 0.24N, 0.30N, 0.24N, 0.11N$	1.09	1.11

is much less if the distribution has multiple modes, particularly if the classes can be centered around the modes. For instance, with cigarette smoking, about $\frac{1}{3}$ or more of the total samples were nonsmokers and it would not be surprising to find peaks around 10 and 20 cigarettes per day. Thus, if we formed two classes of smokers by amount smoked and a third class of nonsmokers, a rough calculation suggests that $V(\hat{\beta}_3)/V(\hat{\beta})$ is around 1.10–1.15 rather than 1.23 as shown in Table 3.4.1.

In practice the reasons for using a few classes at different levels in preference to a full regression analysis have seldom been explicitly stated by investigators. One may guess the reasons as being either (1) for simplicity in the measurement problem and in analysis and presentation, or (2) in some cases, for a judgment that if the amount a_i received by the ith person can be measured only roughly, full regression analysis would not gain much in precision over analysis by use of a few classes. This question has not been fully explored, but to a first approximation it looks as if the presence of errors of measurement in the a_i hurts both the regression and the classification methods to about the same extent, so that the ratios $V(\hat{\beta}_g)/V(\hat{\beta})$ do not change much.

3.5   OTHER POINTS RELATED TO TREATMENTS

This section considers a few miscellaneous points related to the definition of the treatments. Often the level of the treatment varies from person to person because use of the treatment is to some extent voluntary. A well-known example is the distribution of a contraceptive device to married women in a program in which it is planned to estimate the effectiveness of the program. In fact, although the women agree to cooperate, some women in the study sample may not use the device at all, some may use it minimally and inconsistently or use it for a time and then cease, and some may use it as recommended by the planners. From the viewpoint of public policy the main interest may lie in the subsequent birth rate for the sample as a whole, but it is obviously relevant to know to what extent the contraceptive device was actually used, to try to learn the reasons for limited use, and to obtain the birth rate for both the consistent and the inconsistent users. The plans for the study should include regular monitoring of the women in the study sample in order to obtain this information. The studies on smoking were careful when gathering data to distinguish between those who had never smoked and the ex-smokers (those who were not smoking at the time of receipt of the questionnaires, but who had smoked in the past). Comparison of death rates for ex-smokers with death rates for nonsmokers and current smokers, provided much useful information.

The description of the treatment may require setting up a special record-keeping system during the course of the study. To evaluate the worth of making available a limited amount of psychiatric guidance (psychiatric social workers in addition to some expert psychiatric counseling) to a number of families in a health plan, the investigator needs to know with which family members the guidance staff worked, what kinds of mental problems were revealed, the amount of guidance that was given, and so forth. Construction of a good record-keeping system for such purposes must be part of the initial plans.

Sometimes the treatment combines several agents. For example, educational campaign to encourage voter registration or inoculation of children might include announcements on radio and television, distribution of leaflets to houses, and some public lectures. The planning group will want to consider whether they can learn something about the part played by different aspects of the treatment, though this is usually difficult. At least, questions could be framed to ask whether people had been reached by different components, what they remember of the contents of these components, and what they think influenced them in one way or another.

The treatment may take a different form for some people in the study sample than for others. In a study of the effectiveness of wearing seat belts, some people involved in auto accidents may have been wearing both the lap belt and the over-the-shoulder–type belt. Head injuries to children at birth were found to vary in type. The type should obviously be recorded. Existence of more than one type raises the question: Shall we attempt to measure the effects of each variant? A relevant statistical aspect of this question is the following. If alternative forms of the treatment have effects in the same direction, a much larger sample size is needed to distinguish clearly between the sizes of the effects for different types than to measure the overall effect of the treatment, especially if some forms are used by only a small minority. The situation is discussed more specifically in Section 4.1 on sample size.

The best decision is a matter of judgment in a situation like this. If a treatment is used in two forms, some investigators prefer to include both forms, even if the sample size permits only a very imprecise comparison between the effects of the two forms. They argue that they will measure the overall effect of the two forms as they are used, and can at least check that the two forms do not have widely different effects. Others feel that if their sample size is such that the difference between the effects of the two forms is almost certain to be nonsignificant, a report of this finding may lead readers to assume that the two forms have been shown to have equivalent effects, and discourage further research on the minority form. They prefer a "clean" study restricted to the majority form only.

3.6 CONTROL TREATMENTS

Often the investigator first selects the treated group. In fact, an imaginative investigator may notice the existence of an interesting treated group, which gives rise to the study. The investigator then looks for a suitable group on whom this treatment is not acting. Such a group is quite commonly called a "control group," though some investigators prefer the more general term "comparison group," perhaps because the word "control" has some of the features of an advertising slogan, hinting at more power to remove biases than is usually possessed.

Ideally, the requirement for a control group is that it should differ from the treatment group only in that the treatment is absent. The choice of the control group should lead us to expect that if the treatment has no effect, the responses y should have the same mean and shape of distribution in the control group as in the treated group. In particular, the control group should be subject to any selective forces that are known to affect the treatment group and are not themselves possible consequences of the treatment. This last condition can be important. In a study of the effect of birth-related head injuries on children's performance in school at age 11, it would be unwise to have an uninjured control group deliberately chosen so that performance in class at age 6 was similar to that of the treatment group at age 6, unless it were known with certainty that birth-related head injuries had no effect on performance at age 6. Otherwise, a comparison between the treated and control groups at age 11 might have removed part or even all of the treatment effect.

The most common difficulty in the search for a control is that we cannot find a group known to be similar in all other respects to the treated group, particularly when use of the treatment is to some extent voluntary. When making a choice, it is advisable to list the apparent deficiencies of any proposed controls and to try to judge which control seems least likely to produce a major bias in the treated–control comparison. If no single control free from the danger of serious bias can be located, there is merit in having more than one control, particularly if the different controls are suspected of being vulnerable to different possible sources of bias. In this event, a finding that the treated group differs from all of the control groups in the same direction strengthens any claim that there is a real effect of the treatment.

Another requirement sometimes overlooked for a control is that the quality of measurement should be the same in treated and control groups. In a study of tuberculous and nontuberculous (control) families, the same caseworker was assigned to take the measurements in both families so as to guarantee similar quality of measurement. She warned, however, that her measurement would be both more complete and more accurate in the tuberculous families, with whom she had worked for years, than with the control families who were specially recruited for this study and did not know her.

Some investigators like to use data taken from published statistics for the general public as a control, possibly because the general public represents the target population to which they would like the results to apply. Such controls are seldom satisfactory. Many treated groups that appear in studies have some special features apart from the treatment that make them not comparable to the general public. Further, with routinely gathered statistics the meaning of the measurements and their completeness and quality are often quite different from those that apply in a carefully planned study.

To carry this point further, an investigator may claim that a certain treatment is deleterious to health in males because the sickness rate in the treated group is say 20% higher than in the control group, a significant difference statistically. An opponent may state that this claim is unjustified because the sickness rates in the investigator's treated and control groups are *both* definitely lower than those for males with the same age distribution in the general public. The investigator would then retort that the treated group of men differed from men in the general public in certain known respects, that the control group was carefully chosen to differ in the same respects in order to be comparable with the treated group, and that comparison with data for the general public is logically irrelevant.

The investigator is correct, but the question arises: To what target population does the 20% increase in sickness rates apply? The investigator may reply that it applies to the kind of population of which the treated and control groups can be considered a random sample; but as mentioned in Chapter 2, this population may be of no particular interest from the viewpoint of public policy because of the selective forces that affect it, and may even be difficult to envisage. If the investigator claims that the 20% difference is applicable to a more general population of males, he is making a claim without producing supporting data, particularly if this is the first study of this treatment. This point affects a great many observational studies, because the groups that get studied are often specialized in several respects, for quite sound reasons. The same remarks apply to any statement of confidence limits, which account for only the amount of variation in the sampled population.

As studies in different locations by different workers accumulate, there is an opportunity to examine whether this 20% increase is also found in different populations. This is what happens in research on many problems, though often in a haphazard and informal way. The author of the study believes that although the levels of performance in other populations may differ, the size of treatment effect, difference or percent, will apply to other populations. This hope has often, but by no means always, been realized in the past as we mention in the next section.

3.7 THE RESPONSES

The choice of response measurements (measurements to be used to compare the performances under different treatments) is obviously important. In selecting the responses, several points should be considered. Relevance to the stated objectives of the study is an obvious one. To cite an example, reported by Yates (1968), measures of the percentages of buildings destroyed, obtained at great labor from aerial photographs, were used to assess the effects of bombing raids on German cities on Germany's industrial production. For the stated objective this response variable was not particularly relevant, since early British bombing raids concentrated on the town centers, whereas the factories were mainly on the outskirts. A similar statistic constructed for factories gave quite different results and, in order to approach the objective more closely, could be combined with British data on the relationship between damage to factory buildings and actual decrease in factory production.

Different aspects of the responses may be relevant. In the above example, in considering measures of morale in studies of the effects of the bombing raids on morale, one proposed indicator was the proportion of daily absences from work without an obviously operative reason. Another approach involved using a battery of questions about the respondent's opinion on the state of the war, feelings of ability to cope with day-to-day problems, attitudes toward government, and so forth. Either or both types of approach might be advisable, depending on the study's range of objectives. In a study with an obvious primary response measurement, it is worth asking: Are there other aspects of the possible effect of the treatment that should be measured?

The same remark applies to subsidiary measurements that may provide insight on how a causal effect is produced or may strengthen or weaken the evidence that there is a causal effect. For example, an immunization campaign against diphtheria in young children in Britain in the 1940s was followed by scattered reports of paralytic poliomyelitis in some children. At that time there was no polio vaccine.

These reports might suggest that the usual inoculations in children increased their risk of subsequently contracting polio; yet, since many children had been inoculated, the appearance of *some* polio cases among the inoculated was not of itself surprising in the absence of any standard of comparison. Accordingly, in a study conducted in 1949, Hill and

Knowelden (1950) obtained two samples each of 164 children; one sample was of children with polio, and the control sample was of children without polio. The samples were matched for age, sex, and place of birth. The aim was to see whether a much higher proportion of the polio cases than the control sample had been inoculated. The investigators found that 96 of the polio cases and 83 of the control samples had received inoculations—a difference that was not striking and certainly not statistically significant.

As additional information, the investigators recorded (1) the dates of any inoculations, and (2) the sites (left arm, left leg, etc.) of the inoculations. This information enabled two further comparisons to be made. Of 17 inoculations made during the month immediately preceding the polio attack, 16 occurred in polio cases and 1 in the control sample. Of the inoculations made more than one month previous, 80 occurred in polio cases and 82 in the control sample. A second piece of information was provided by the polio cases alone. For inoculations made *more* than one month previous to the onset of paralysis, the site of the inoculation injection was also one of the sites of paralysis in 13 out of 65 cases, or 20%. For inoculations *less* than one month previous, the corresponding figures were 29 out of 36 cases, or 81%. These two results both pointed to an increased risk of paralytic polio from inoculations in the month preceding the time when polio becomes epidemic each year. Taken together with other data, these results led to a recommendation that doctors should avoid giving standard inoculations to children during this period.

The investigator should of course be aware of what is known about the quality of any proposed measuring process. Often, this aspect presents no problem. At the other extreme, there are instances in which a satisfactory method has not yet been developed for measuring a given type of response; therefore, sound research studies cannot proceed until some breakthrough in measurement has occurred. In such cases a common situation is that several different measuring techniques (e.g., by a battery of questions) have been developed, but it is not clear exactly what is being measured by each technique. If information on the agreement between different techniques is scanty, use of two of them on each person in a study is worth considering as a means of picking up useful comparative information for future studies.

Sometimes the process of measurement requires use of several similar instruments (e.g., interviewers, judges, raters, laboratories). A standard precaution is to have each judge measure the same proportion of subjects, preferably selected at random, from each treatment group. Whenever feasible, it is also worth the inconvenience to ensure that each judge is ignorant of the treatment group to which any subject belongs at the time when the judge is rating the subject. Otherwise, consistent differences between judges in levels of rating, or the judge's preconceived ideas of how the treatments should rank in performance, may produce fallacious differences between treatments. These precautions are easily overlooked, as experience has shown, in a study that involves numerous detailed operating decisions.

The size and range of the study may determine the type and, therefore, the quality of the measuring instrument for responses. For instance, in a study of nonhospitalized mental ill-health, the choice might lie, depending on its size, between measurement by trained psychiatrists, by psychiatric nurses, or by a standard questionnaire administered by lay interviewers. In a large-sample study of school education, information about schoolwork conditions in the home and the parent's attitudes and aspirations with respect to their children's education, might have to be obtained from questions answered by the children in school because the available resources do not permit use of direct interviewing of parents in their homes.

It is difficult to advise to what extent the original aims and scope of a study should be deliberately restricted by sacrificing some of the original objectives in order to permit a higher quality of measurement. The weaknesses of the smaller-sample study are likely to be restricted range of questions, reduced precision, and limitation to a subpopulation much narrower than the population of interest. There is merit in a restricted study that can (1) indicate whether the case for an extensive study is strong or weak, (2) allow internal comparisons between the best-available measuring techniques and less-expensive ones that would have to be used in a broad study, and (3) provide valuable experience in the problems of conducting this type of study. The weaknesses of a large extensive study in which the investigators bit off more than they can chew are likely to be high nonresponse rates and measurements clearly vulnerable to bias, with the consequence that the main conclusions are subject to serious question. Its strength may be that it can attempt to sample the population of interest. Another weakness of the large study is that it may fall of its own weight and never be completed. Fortunately, these often die aborning.

In practice, the decision for or against a large study will be influenced also by the amount of public interest in the problem, the pressure for early results, and scale of operation that attracts the principal investigators.

3.8 TIMING OF MEASUREMENTS

Problems of timing the study measurements arise if the treatment is expected to have responses that last for some time into the future. Examples are the fluoridation of a town's water or a rise in the general tax rate.

Shortly before the treatment is applied, a relevant response variable is measured on a sample of children or families, for instance a survey of the numbers of decayed, missing, and filled teeth by age and sex, or of family spending and saving. The following question then arises: How long will these response variables be measured after the start of the treatment? If little is known in advance about the shape of the time–response curve, no definite answers can be given to this question, except perhaps to schedule at least two subsequent observations, to keep later study plans flexible, and to speculate about the likely nature of the time–response curve from a combination of theoretical ideas on the nature of the presumed causal process and of any related data from other studies.

Usually these first two measurements have to be scheduled, at least tentatively, at the time when the study is being planned and funds allotted or sought. When these measurements have been obtained, a mathematical model of the nature of the response curve may help in deciding on the desirability and timing of any subsequent measurements. Suppose that the response is expected to increase with time until it reaches some steady maximum value or asymptote. A response curve of this type is often well enough represented by the curve $Y = \mu_0 + \alpha(1 - e^{-\beta t})$, where μ_0 is the initial level at time $t = 0$. When t is large, $e^{-\beta t}$ becomes negligible, so that the parameter α represents the maximum increase. The parameter β indicates how quickly this maximum is reached. At time $t = 3/\beta$, for instance, the curve is within 5% of the maximum increase. If β is doubled, this 95% of the total increase is reached in half the time.

We assume a single treated group. From the initial sample mean and the means $\bar{y}_1$ and $\bar{y}_2$ of the two later responses, we can estimate μ, α, and β. From these estimates and the equation of the curve we can, in turn, estimate (1) how much benefit would be obtained from one or two additional future observations, and (2) the best later times at which to take these observations, in order to obtain as precise estimates as possible of α, β, or the course of the response curve. This use of theory might be particularly helpful if the calculations revealed that the first two posttreatment measurements had been taken much too soon, as might happen with a treatment whose effects are slow to appear.

If we had initially a rough idea (perhaps from other studies) of the time taken to reach 95% of the increase, from which the value of β could be estimated, this information would be useful in planning the times τ and 2τ of the first two posttreatment observations. For sketching the course of the curve, a good compromise is to take τ as the time when the increase has reached about 70% of its maximum. Taking τ at 50% of the maximum is a bit early for this curve.

3.9 SUMMARY

This chapter deals with preliminary aspects of studies intended to compare a limited number of groups of individuals. The groups are exposed to different agents or experience, called *treatments*, whose effects it is desired to study.

An initial written statement of the objectives of the study, with reasons for choosing these objectives, is essential. This may be in the form of questions to be answered, hypotheses to be tested, or effects to be estimated. This statement is helpful in selecting the locale of the study, the specific groups to be studied, the data needed to describe the treatments, the response variables, and the kind of statistical analysis required.

In exploratory studies there is a tendency to think mainly in terms of tests of significance, because the investigator feels that an objective of finding out whether there is *some* effect, and in which direction, is as much as can be accomplished. Thinking in terms of estimating the size of the effect and its practical importance may make the investigator more aware of potential sources of bias and may lead to a more informative statistical analysis.

When the level of treatment varies from group to group, a two-group study with levels widely apart is usually advisable in small initial studies. A third group at an intermediate level adds little or nothing to the estimation of a linear effect, but may either provide a test of linearity or serve as some reassurance about bias if the response is thought to be linear.

When the level of treatment varies from person to person and can be measured at least roughly for individual persons, a common practice is to divide the persons into two or more groups according to these individual levels. For the estimation of a linear effect of levels, the precision obtained with different numbers of groups, the best group sizes, and the relative precision given by groups of equal size are presented.

Special procedures or record keeping may be required when the treatment is complex or when exposure to it is to some extent voluntary. Issues relevant to the selection of an untreated comparison group, sometimes called a control group, are discussed.

In selecting measurements to be made of the responses in groups under different treatments, relevance to the objectives of the study is an obvious criterion. Another is the quality of the measurement. Different aspects of the responses, measurable by different variables, may deserve study. Subsidiary measurements may clarify the interpretation of the main results.

If the process of measurement requires several instruments of the same type (e.g., interviewers, judges, raters, laboratories), standard precautions are to have each judge measure the same proportion of subjects from each treatment group and to conceal from any judge (whenever feasible) the particular group that he is measuring at any specific time.

Sometimes a treatment, such as fluoridation of a town's water or a rise in taxes, will be applied for a relatively long time. Decisions are required about the times at which the treated group is to be observed. If the general shape of the likely time–response curve can be envisaged, statistical theory may assist these decisions.

REFERENCES

Buck, A. A., et. al. (1968). Coca chewing and health. *Am. J. Epidemiol.*, **88**, 159–177.

Cochran, W. G. (1968). The effectiveness of adjustment by subclassification in removing bias in observational studies. *Biometrics*, **24**, 295–313 [Collected Works #90].

Cox, D. R. (1957). Note on grouping. *J. Am. Stat. Assoc.*, **52**, 543–547.

Hill, A. B. and J. Knowelden (1950). Inoculation and poliomyelitis. *Br. Med. J.* ii, 1–16.

Seltser, R. and P. Sartwell (1965). The influence of occupational exposure to radiation on the mortality of American radiologists and other medical specialists. *Am. J. of Epidemiol.*, **81**, 2–22.

Yates, F. (1968). Theory and practice in statistics, *J. Roy. Statist. Soc.*, Ser. A, **131**, 463–477.

CHAPTER 4

Further Aspects
of Planning

4.1 SAMPLE SIZE IN RELATION TO TESTS OF SIGNIFICANCE

Suppose that the investigator has identified two or more groups of subjects whose mean values for some response variable y he wishes to compare. A decision must be made about the size of sample to be selected from each group. Sometimes this decision is controlled largely by cost or availability considerations. A group that is of particular interest may contain only 80 subjects, or the budget may limit the study to two samples of sizes not exceeding 200 each. In the absence of such limitations, statistical theory provides certain formulas as a guide in making decisions about sample size. Use of these formulas may present difficulties, either because the formulas oversimplify the actual conditions of the survey or because the investigator does not have certain information about the study that the formulas require. Nevertheless, it is worth finding out what light these formulas throw on the sample-size issue even when the size is limited by costs or availability.

Calculation of sample size in relation to a test of significance is most often made in exploratory studies. Suppose that there are two groups of subjects exposed to different agents, or one group exposed to an agent and a control group unexposed. If the study fails to find a significant difference $\bar{y}_1 - \bar{y}_2$, the investigator knows that he will have obtained an inconclusive result. The group means have not been shown to be different, but neither have they been shown to be essentially the same, since this conclusion would amount to assuming that the null hypothesis has been proved correct, or nearly correct.

If $\bar{d} = \bar{y}_1 - \bar{y}_2$ and δ is the unknown population difference between the group means, the investigator might reason that he does not mind finding $\bar{d}$

50

nonsignificant if δ is small. However, if δ is large enough to be of practical importance, the investigator wants to have a high probability of detecting that there is a difference by finding $\bar{d}$ significant. Such a result may encourage later work that will estimate δ more accurately. This leads to the question: For a given δ and given sample sizes from the two groups, what is the probability of obtaining a significant $\bar{d}$? This probability is called *the power of the test*.

This calculation is easily made if bias can be ignored and if we can assume $\bar{d}$ normally distributed with mean δ and standard deviation $\sigma_{\bar{d}}$. Let z_α be a nonnegative number such that the probability that a normal deviate exceeds z_α is α. For example, if $z_\alpha = 0$, $\alpha = 0.5$ and if $z_\alpha = 1.96$, then $\alpha = 0.025$, since we are considering only the right-hand tail of the normal distribution.

We can now calculate the probability that $\bar{d}$ is significant. We start with a one-tailed test (δ assumed $\geqslant 0$), since this is slightly easier. Clearly, $\bar{d}$ is significant if it exceeds $z_\alpha \sigma_{\bar{d}}$. Thus, we want to find the probability that $\bar{d}$ exceeds $z_\alpha \sigma_{\bar{d}}$. If $\bar{d}$ is normally distributed with mean δ and standard deviation $\sigma_{\bar{d}}$, the standard normal deviate corresponding to $\bar{d}$ is therefore $(\bar{d} - \delta)/\sigma_{\bar{d}}$. Now setting $\bar{d} = z_\alpha \sigma_{\bar{d}}$, the threshold significant value, we compute the probability that $\bar{d}$ is significant by calculating

$$z = \frac{z_\alpha \sigma_{\bar{d}} - \delta}{\sigma_{\bar{d}}} = z_\alpha - \frac{\delta}{\sigma_{\bar{d}}} \qquad (4.1.1)$$

and reading the probability that a normal deviate exceeds z. If we do this for fixed n and α and various δ, we produce a curve called *the power function* which relates power and δ.

If the study is planned to have two independent groups, each of n subjects, then $\sigma_{\bar{d}} = \sqrt{2}\,\sigma/\sqrt{n}$, where σ is the standard deviation per subject in both groups. Formula (4.1.1) then becomes

$$z = z_\alpha - \sqrt{\frac{n}{2}}\,\frac{\delta}{\sigma} \qquad (4.1.2)$$

Let us consider some examples which illustrate the use of formulas in the estimation of sample size.

Example 1. Suppose that costs limit the sample sizes to $n = 100$. Related data indicate that σ is about 1. The investigator thinks that if δ is as large as 0.3, this is important enough so that he would like to obtain a significant difference. What is the probability? In this case $z_\alpha = 1.64$ for a one-tailed test at the 5% level. Thus by (4.1.2),

$$z = 1.64 - \sqrt{50}\,(0.3) = 1.64 - 2.12 = -0.48$$

The normal tables give $P = 0.68$ for the probability of exceeding z, a little disappointing, but as good as many studies can offer.

In a two-tailed test, $\bar{d}$ can be either significantly positive or significantly negative. In our notation the conditions are $\bar{d} > z_{\alpha/2}\sigma_{\bar{d}}$ or $\bar{d} < -z_{\alpha/2}\sigma_{\bar{d}}$. Note the subscript $\alpha/2$; if the two-tailed probability is to be 0.05, the one-tailed probability must be 0.025. If $\delta > 0$, a verdict that $\bar{d} < -z_{\alpha/2}\sigma_{\bar{d}}$ would be a horrible mistake, since we find $\bar{d}$ significant, but in the wrong direction. Fortunately, if the probability that $\bar{d}$ is significant in the correct direction is at all sizable (e.g., > 0.2), the probability that $\bar{d}$ is significant in the wrong direction is tiny and can be ignored. Hence in a two-tailed test we can calculate the probability that $\bar{d}$ is significant and in the correct direction by amending (4.1.1) to the probability that a normal deviate exceeds

$$z = z_{\alpha/2} - \frac{\delta}{\sigma_{\bar{d}}} \qquad (4.1.3)$$

With two planned samples each of size n, we can now calculate the needed value of n such that the probability of finding a significant $\bar{d}$ has any desired value β. Take $\beta > 0.5$, since we want the probability to be high. Earlier, we defined z_{α} $(\geqslant 0)$ as a value for which the probability that a normal deviate exceeds z_{α} is α. This definition restricts us to values of $\alpha \leqslant 0.5$. If $\beta > 0.5$, the value of z such that the probability β of exceeding this z is $-z_{(1-\beta)}$. By the symmetry of the normal curve, the probability that $z \leqslant -z_{(1-\beta)}$ is $(1 - \beta)$, so that the probability that $z > -z_{(1-\beta)}$ is β for $\beta > 0.5$.

To summarize, if we want a *one-tailed* test to have probability β (> 0.5) of finding a significant result at level α, we write

$$-z_{1-\beta} = z_{\alpha} - \sqrt{\frac{n}{2}}\,\frac{\delta}{\sigma} \qquad (4.1.4)$$

and solve for n, giving

$$n = \frac{2(z_{\alpha} + z_{1-\beta})^2 \sigma^2}{\delta^2} \qquad (4.1.5)$$

If the test is *two-tailed*,

$$n = \frac{2(z_{\alpha/2} + z_{1-\beta})^2 \sigma^2}{\delta^2} \qquad (4.1.6)$$

For one- and two-tailed tests, Table 4.1.1 shows the multipliers of σ^2/δ^2 for specified probabilities, from 0.5 to 0.95, of finding a significant difference.

As before, the ratio δ/σ that is of importance must be specified in order to use Table 4.1.1.

Example 2.   A pilot study suggests that σ may be about 6 while the investigator would like $\beta = 0.95$ if δ is 2 in a two-tailed test at the 5% level. For this, $\sigma^2/\delta^2 = 9$ and $n = (26.6)(9) = 239$ in each sample.

Table 4.1.1 may also be used as an approximation when the response is a binomial proportion and we are comparing independent samples from two populations whose proportions are p_1 and p_2. The numerical factors in Table 4.1.1 remain the same, but σ^2/δ^2 is replaced by

$$\frac{p_1 q_1 + p_2 q_2}{2(p_1 - p_2)^2} \tag{4.1.7}$$

with $q = 1 - p$, or $q = 100 - p$ if p is expressed as a percentage. If the first population is a control or a standard method, p_1 may be known fairly well from previous studies. The value of p_2 has to be inserted from consideration of the size of difference $|p_1 - p_2|$ that the investigator does not want to "miss" in the sense of this test of significance.

Example 3.   If $p_1 = 6\%$, $p_2 = 3\%$, then $q_1 = 94\%$, $q_2 = 97\%$. To have an 80% chance of finding a significant difference in a one-tailed test, we require

Table 4.1.1.   Multipliers of σ^2 / δ^2 Needed to Give n for a Specified Probability of Finding a 5% Significant Difference in the Correct Direction

Probability	One-Tailed Test	Two-Tailed Test
0.5	5.4	8.0
0.6	7.2	10.2
0.7	9.4	12.7
0.8	12.4	16.2
0.9	17.1	21.5
0.95	21.6	26.6

the size of each sample to be

$$n = \frac{(12.4)[(6)(94) + (3)(97)]}{(2)(9)} = 589$$

In this and the preceding section the sample-size formulas that involve n assume two *independent* samples. Often, the samples are matched or paired by certain characteristics of the subjects. If so, the quantity σ^2/δ^2 in Table 4.1.1 is replaced by $\sigma^2(1 - \rho)/\delta^2$, where ρ is the correlation coefficient between members of the same pair. Sometimes ρ can be guessed if an estimate of σ^2 is available. Alternatively, if $d_j = y_{1j} - y_{2j}$, the difference between the members of the jth pair, σ^2 may be replaced by $\sigma_d^2/2$. If σ_d^2 is being estimated from a past study, this should of course have employed the same criteria for matching. With binomial data, a fair amount of evidence suggests that pairing is usually only moderately effective in increasing precision. Calculation of n by formula (4.1.6) for independent samples will be on the conservative side, but not badly wrong.

The method extends to cases not so simple as two samples of size n, provided that $\sigma_{\bar{d}}$ can be calculated. We give two numerical illustrations in Example 4 and a general algebraic one in Example 5.

Example 4. The group on which a treatment acts will provide only 50 subjects. The control group is not so restricted. The investigator guesses that the probability of detecting his desired δ will not be high if only 50 control subjects are used. How much better does the investigator do if n is 100 or 200 for the control sample? We have $\sigma = 10$, $\delta = 4$. We revert to formula (4.1.3), with $z_{\alpha/2} = 1.96$ in place of z_α since a two-tailed test is desired.

$$z = 1.96 - \frac{4}{\sigma_{\bar{d}}}$$

With 50 control subjects, $\sigma_{\bar{d}} = \sqrt{2}\,\sigma/\sqrt{50} = 2$, so that $z = -0.04$, giving a probability 0.52. With 100 and 200 controls,

$$\sigma_{\bar{d}} = \sigma\sqrt{\frac{1}{50} + \frac{1}{100}} = 1.732; \qquad \sigma_{\bar{d}} = \sigma\sqrt{\frac{1}{50} + \frac{1}{200}} = 1.581$$

so that $z = -0.35$ and -0.57, with probabilities 0.64 and 0.72, respectively.

Example 5. In what is called a before–after study, measurements are taken in each group both before an agent has been applied to the group and at some time afterwards. The quantity of interest is often

$$\bar{d} = (\bar{y}_{1a} - \bar{y}_{1b}) - (\bar{y}_{2a} - \bar{y}_{2b})$$

the mean difference in the changes associated with each agent. If each measurement has the same variance σ^2, and the correlation between before and after measurements in the same group is ρ, then

$$\bar{y}_{1a} - \bar{y}_{1b} = \bar{d}_1 \quad \text{and} \quad \bar{y}_{2a} - \bar{y}_{2b} = \bar{d}_2$$

$$\text{Var } \bar{d}_1 = \frac{\sigma^2 - 2\rho\sigma^2 + \sigma^2}{n} = \frac{2\sigma^2}{n}(1 - \rho) = \text{Var } \bar{d}_2$$

Combining this information, we find

$$\sigma_{\bar{d}} = \sqrt{\sigma_{\bar{d}_1}^2 + \sigma_{\bar{d}_2}^2} = \sigma\sqrt{\frac{4}{n}(1 - \rho)}$$

This $\sigma_{\bar{d}}$ is used in formula (4.1.1).

4.2   SAMPLE SIZE FOR ESTIMATION

Sometimes investigators prefer to look at sample-size formulas from the viewpoint of closeness of estimation rather than of testing significance. This is so, for instance, if there is a good deal of presumptive evidence in advance that a treatment will produce some effect; the question is whether our estimates will be sufficiently accurate as a basis for action. As before, we assume that bias is negligible and that our estimated difference $\bar{d}$ can be taken to be normally distributed. Thus, if δ is the population difference, $\bar{d}$ should lie within the limits $\delta \pm L$ with about 95% probability where

$$L = 2\sigma_{\bar{d}} \tag{4.2.1}$$

In the simplest application to two independent samples each of size n

$$L = \frac{2\sqrt{2}\,\sigma}{\sqrt{n}} = \frac{2.82\sigma}{\sqrt{n}} \tag{4.2.2}$$

where σ is the within-group standard deviation. Formulas (4.2.1) and (4.2.2) can be used in a number of ways.

Let us consider examples which illustrate the use of formulas to produce the sample-size estimate correct to a certain limit with high probability.

Example 1. If $\bar{d}$ is desired to be correct to within specified limits of error $\pm L$ (apart from a 1 in 20 chance), we have from (4.2.2)

$$n = \frac{8\sigma^2}{L^2} \tag{4.2.3}$$

as the size of each group.

Example 2. Suppose that financial or other considerations limit n to 400 and that σ is thought to be about 3. From (4.2.2)

$$L = \frac{2.82\sigma}{\sqrt{n}} = 0.423$$

Consideration as to whether this is satisfactory will probably involve some thought about any action to be taken. The situation might be "I feel that action is necessary if $\delta \geqslant 1$ and will argue for action if $\bar{d} \geqslant 1$, but not otherwise." Clearly, if we are unlucky we may find ourselves arguing for action if δ is only $(1 - 0.423) = 0.577$, or failing to argue for action when δ is nearly as high as 1.423. The issue then depends on whether mistakes of this kind are regarded as tolerable.

Example 3. If the response variable is a binomial proportion so that $\bar{d} = \hat{p}_1 - \hat{p}_2$, then with independent samples, $\sqrt{2}\,\sigma$ becomes $\sqrt{p_1 q_1 + p_2 q_2}$. Hence (4.2.2) becomes

$$L = \frac{2\sqrt{p_1 q_1 + p_2 q_2}}{\sqrt{n}} \tag{4.2.4}$$

This formula holds whether p_1, p_2, and L are all expressed in proportions or percentages. In proportions, $q_i = 1 - p_i$; in percentages, $q_i = 100 - p_i$.

Suppose samples of $n = 3600$ can be run and the failure rate (response) in the control group is 10%. The failure rate in the treatment group is not known, but if it is as low as 5%, how well is the improvement in failure rate estimated? With $p_1 = 10$, $p_2 = 5$,

$$L = \tfrac{2}{60}\sqrt{(10)(90) + (5)(95)} = 1.24\%$$

This might seem good enough. Even if the treatment is ineffective, $p_2 = 10\%$, we have

$$L = \tfrac{2}{60}\sqrt{(10)(90) + (10)(90)} = 1.41\%$$

so that it is unlikely that p_2 would be estimated as more than 1.41% lower than p_1.

Example 4. When estimating a treatment effect from two groups of subjects, the investigator may have some subgroups for which it would be informative (1) to estimate the treatment effect separately in each subgroup, and (2) to compare the sizes of the treatment effects for different subgroups. Unfortunately, as is well known, much larger sample sizes are needed for case (1) and particularly for case (2) than for the estimation of an overall treatment effect.

Suppose for illustration two subgroups contain proportions ϕ and $1 - \phi$ of the subjects in each of a treatment and a control population. With male–female subgroups, ϕ might be about 0.5; with white–black subgroups, ϕ might be 0.8 or 0.9. The values of $\sigma_{\bar{d}}$ are therefore $\sqrt{2}\,\sigma/\sqrt{n}$ for the overall effect, approximately $\sqrt{2}\,\sigma/\sqrt{\phi n}$ and $\sqrt{2}\,\sigma/\sqrt{(1-\phi)n}$ for the individual subgroup effects, and $\sqrt{2}\,\sigma/\sqrt{\phi(1-\phi)n}$ for the difference in effect from one subgroup to the other. Since $L = 2\sigma_{\bar{d}}$, the multipliers of σ^2/L^2 required to find n are 8, $8/\phi$, $8/(1-\phi)$, and $8/\phi(1-\phi)$. These multipliers show the relative sample sizes needed to attain the same error limit $\pm L$, and are presented in the list below for $\phi = 0.5$ and 0.9.

ϕ	Overall Effect	Effect in Subgroup 1	Subgroup 2	Difference in Effects
0.5	8	16	16	32
0.9	8	9	80	89

With subgroups of equal size, the most favorable case, estimation of effects separately in each subgroup requires twice the sample size, while estimating the difference in effects requires four times the size. With k equal subgroups the multipliers are k and $2k$. As the case $\phi = 0.9$ illustrates, the situation is much worse when some subgroups are relatively small.

These results are not intended to deter an investigator from examining effects separately in different subgroups. But the accompanying standards of precision are lower, and large samples are usually needed to estimate a difference in effects from one subgroup to another.

Example 5. If the cost of sampling and measurement is much cheaper in population 1 than in population 2, the question is occasionally asked: What sample sizes n_1 and n_2 will provide a specified value of $V(\bar{d})$ at minimum cost? Let

$$\text{cost} = C = c_1 n_1 + c_2 n_2 \ (c_1 < c_2); \qquad V = V(\bar{d}) = \frac{\sigma_1^2}{n_1} + \frac{\sigma_2^2}{n_2}$$

The calculus minimum-cost solution is

$$\frac{n_1}{\sigma_1\sqrt{c_2}} = \frac{n_2}{\sigma_2\sqrt{c_1}} = \frac{n}{\sigma_1\sqrt{c_2} + \sigma_2\sqrt{c_1}}; \qquad n = \frac{\left(\sigma_1\sqrt{c_2} + \sigma_2\sqrt{c_1}\right)^2}{V\sqrt{c_1 c_2}}$$

Assuming σ_1 and σ_2 are roughly equal, we have $n_1/n_2 = \sqrt{c_2}/\sqrt{c_1}$. Unless the cost-ratio c_2/c_1 is extreme, however, the saving over equal sample sizes is modest, since

$$\frac{C_{equal}}{C_{min}} = \frac{2(c_1 + c_2)}{\left(\sqrt{c_1} + \sqrt{c_2}\right)^2}$$

$$= \frac{2(1 + c_2/c_1)}{\left(1 + \sqrt{c_2/c_1}\right)^2}.$$

Equal sample sizes cost only 3% more if $c_2/c_1 = 2$, 11% more if $c_2/c_1 = 4$, and 27% more if $c_2/c_1 = 10$.

4.3   THE EFFECT OF BIAS

As mentioned, the formulas in the preceding sections assume that any bias in the estimates is negligible. The effect of bias on the accuracy of estimation was discussed in Section 2.5. To cite results given there, suppose n has been determined so that the probability is 0.95 that $\bar{d}$ lies in the interval $(\delta - L, \delta + L)$ in the absence of bias. The presence of an unsuspected bias of amount $\leqslant 0.2L$ decreases the 0.95 probability only trivially. If $B = 0.5L$, the 0.95 probability is reduced to 0.84 and to less than 0.50 if B/L exceeds 1.0. For given $f = B/L$ a table can be constructed which shows the amount n must be increased over n_0 in the "no bias" situation in order to keep this probability at 0.95. Table 4.3.1 shows the ratio n/n_0 for $f = 0.2(0.1)0.9$.

It is not likely that Table 4.3.1 can be used for estimating n in planning a specific survey, because of ignorance of the value of f. If an investigator somehow knew f fairly well, he/she would try to adjust $\bar{d}$ in order to remove the bias and would then face a different estimation problem. The table helps, however, in considering a possible trade-off between reduction of bias and reduction of random sources of error. For instance, if by better planning or more-accurate measurements the value of f could be reduced from 0.6 to 0.3, a sample size of $1.40n_0$ would be as effective as one of

Table 4.3.1. Ratio of n/n_0 Needed in Order to Give 95% Probability that $\bar{d}$ is Correct to Within $\delta \pm L$ When Bias of Amount fL is Present. (Note that n_0 is the sample size in the "no bias" case.)

$f = B/L$	n/n_0
0.2	1.13
0.3	1.40
0.4	1.89
0.5	2.72
0.6	4.22
0.7	7.51
0.8	16.9
0.9	67.6

$4.22n_0$, but would be only about one-third the size. A more expensive method of data collection may save money if it reduces the bias sufficiently.

4.4 MORE COMPLEX COMPARISONS

The illustrations of sample-size problems in Sections 4.1 and 4.2 have referred mainly to the difference between the means of two groups. In studies with more than two groups the comparison of primary interest may be more complex. The procedure here is to define $\bar{d}$ as the estimated comparison of interest, with δ as the population value of the comparison. For a specified probability β (> 0.5) of "detecting" δ, we may rewrite formula (4.1.4) more generally as

$$-z_{(1-\beta)} = z_\alpha - \frac{\delta}{\sigma_{\bar{d}}} \qquad (4.4.1)$$

If we want $|\bar{d} - \delta| \leqslant L$ apart from a 1 in 20 chance, we can use (4.2.1),

$$L = 2\sigma_{\bar{d}} \qquad (4.4.2)$$

From the nature of the comparison we should be able to express $\sigma_{\bar{d}}$ in terms of n, the size of each group, and then solve for n from (4.4.1) or (4.4.2).

The following examples involve studies having more than two groups.

Example 1. A study has three groups, containing amounts 0, 1, and 2 or a, $a + 1$, and $a + 2$ of the treatment. If the response is thought to be linearly related to the amount of the treatment, the quantity of primary interest may be the average change in response per unit increment in amount. For this,

$$\bar{d} = \frac{\bar{y}_3 - \bar{y}_1}{2} \quad \text{and} \quad \sigma_{\bar{d}} = \frac{\sigma}{\sqrt{2n}}$$

With four groups having amounts 0, 1, 2, and 3 or a, $a + 1$, $a + 2$, and $a + 3$ of the treatment, the corresponding estimate of the average change in y per unit increase in amount is

$$\bar{d} = \frac{3\bar{y}_4 + \bar{y}_3 - \bar{y}_2 - 3\bar{y}_1}{2} \quad \text{and} \quad \sigma_{\bar{d}} = \frac{\sigma\sqrt{5}}{\sqrt{n}}$$

More generally, suppose that we have k groups having amounts $x_1, x_2, \ldots, x_k$. We use weights

$$w_i = x_i - \bar{x}$$

and

$$\bar{d} = \sum w_i \bar{y}_i$$

with variance

$$V(\bar{d}) = \frac{\sum w_i^2 \sigma^2}{n} = \frac{\sigma^2}{n} \sum (x_i - \bar{x})^2$$

Example 2. This example is artificial, but illustrates the way in which the formulas are adapted for regression studies. A firm has several factories doing similar work. Certain tasks sometimes done by the workers require a high degree of skill, concentration, and effort, and good performance in these tasks is important. The management finds that one factory offers one unit per hour of extra pay as incentive for this work, another factory offers three units per hour, and a third factory offers no incentive pay.

The management considers taking a random sample of workers in each of these three factories and recording performance scores. It is proposed to estimate the average change in performance per unit extra incentive pay. If the true increase in performance per unit incentive pay is at least 4%, the management would like a 95% chance of declaring the estimated increase to be significant (5% one-tailed test). What sample size is needed in each factory?

Assuming a linear effect of the incentive-pay performance, the values of x are 0, 1, and 3, with $\Sigma(x - \bar{x})^2 = 14/3$. Hence $\sigma_{\bar{d}} = \sigma\sqrt{14/3n}$. From (4.4.1),

$$\sqrt{\frac{3n}{14}}\frac{\delta}{\sigma} = z_\alpha + z_{1-\beta}$$

$$n = \frac{14}{3}(z_\alpha + z_{1-\beta})^2\frac{\sigma^2}{\delta^2}$$

The management wants $\delta = 0.04\mu$, where μ is the performance level under no incentive pay. For a one-tailed 5% test and probability 0.95, z_α and $z_{1-\beta}$ are both 1.64, so that

$$n = \frac{14(3.28)^2}{(3)(0.0016)}\left(\frac{\sigma}{\mu}\right)^2 \approx 31400\left(\frac{\sigma}{\mu}\right)^2.$$

Work performance scores are not kept routinely, but some recent data indicate a between-workers coefficient of variation $(100\ \sigma/\mu)$ of 15% for nonincentive workers. Thus $\sigma/\mu = 0.15$, giving $n = 706.5$ for the sample from each of the three factories. This investigation may be more expensive than the manufacturer is willing to pay. If we reduced the power from 0.95 to 0.5, the sum of the z's would reduce to 1.64, which is half of 3.28. This would reduce the sample sizes to $706/4 \approx 176$, a considerable reduction in effort, but at a large price in ability to detect an improvement.

Some sample-size problems require distributions different from the normal; solutions are sometimes available from results in the literature. For instance, an investigator might be primarily interested in comparing the amounts of variability in two groups as estimated by the sample variances. A rough answer to this problem can be obtained from tables of the F distribution, though this assumes normality in the original distribution of y and the sizes have to be increased substantially if the distribution of y is long-tailed, with positive kurtosis.

4.5 SAMPLES OF CLUSTERS

The illustrations in Sections 4.1 and 4.2 may be unrealistic for a second reason, in that the structure of the sample is more complex than has been assumed. A common case is that in which the individuals in a sample fall naturally into subgroups or clusters, the sample being drawn by clusters.

These clusters might be families, Boy Scout troops, Rotary Clubs, schools, or church congregations.

Sampling now proceeds in two stages. First, a sample of k clusters is chosen at random; then from each chosen cluster a sample of individuals is randomly drawn, n_j of them from cluster j. Letting y_{jr} represent the rth observation in the jth sampled cluster, we fix ideas by writing

$$y_{jr} = \mu + \gamma_j + e_{jr} \qquad \left(j = 1, 2, \ldots, k; r = 1, 2, \ldots, n_j; \qquad \sum_{j=1}^{k} n_j = n \right)$$

(4.5.1)

In Eq. (4.5.1), μ is the population mean. γ_j is the departure from μ of the jth cluster's mean so γ_j is a random quantity with mean zero and a variance that we shall name σ_γ^2, the between-cluster variance. The value e_{jr} is the random departure of y_{jr} from its own cluster mean; thus e_{jr} also has mean zero, and we shall use σ_{ej}^2 for its variance, noting that this within-cluster variance may be different from cluster to cluster.

With clusters of a given type, the sample-size problem is likely to be that of choosing k, the number of clusters in the group that receives a specified treatment. If $n = \Sigma n_j$ is the total number of individuals in the sample, the average size of cluster is $\bar{n} = n/k$. The most-natural estimate is usually the sample mean per individual, $\bar{y} = \Sigma\Sigma y_{jr}/n$. Sometimes, however, it is advantageous to consider another estimate, $\bar{\bar{y}}_c = \Sigma\bar{y}_j./k$, the unweighted mean of the cluster means. From (4.5.1),

$$\bar{\bar{y}}_c = \mu + \frac{1}{k}\sum \gamma_j + \frac{1}{k}\sum \bar{e}_j.$$

and

$$V(\bar{\bar{y}}_c) = \frac{\sigma_\gamma^2}{k} + \frac{1}{k^2}\sum \frac{\sigma_{ei}^2}{n_j} = \frac{1}{k}\left(\sigma_\gamma^2 + \frac{\sigma_e^2}{\bar{n}_h} \right)$$

(4.5.2)

if $\sigma_{ej}^2 = \sigma_e^2$ is a constant, where $\bar{n}_h$ is the harmonic mean of the n_j. A property of $\bar{\bar{y}}_c$ is that the sample variance between cluster means provides an unbiased estimate of $V(\bar{\bar{y}}_c)$, namely,

$$V(\bar{\bar{y}}_c) = \sum \frac{\left(\bar{y}_j. - \bar{\bar{y}}_c \right)^2}{k(k-1)} = \frac{\hat{\sigma}_b^2}{k}$$

This estimate, with $(k-1)$ degrees of freedom, is unbiased regardless of whether the within-cluster variances σ_{ej}^2 vary from cluster to cluster. Thus the simple formulas of the preceding section may be used in estimating the number of clusters needed, with $\hat{\sigma}_b^2$ replacing σ^2 and k replacing the previous n. This would of course require previous data for the same type of cluster.

The result about $V(\bar{\bar{y}}_c)$ also holds when the response is a $0-1$ variate. If the estimate is the mean $\bar{\bar{p}}_c$ of the cluster mean proportions $\bar{p}_j$, the quantity $\Sigma(\bar{p}_j - \bar{\bar{p}}_c)^2 / k(k-1)$ is an unbiased estimate of $V(\bar{\bar{p}}_c)$, replacing the familiar $\bar{\bar{p}}\bar{\bar{q}}/n$.

With the ordinary sample mean per person,

$$\bar{\bar{y}} = \frac{1}{n}\sum_j \sum_r y_{jr} = \mu + \frac{1}{n}\sum n_j \gamma_j + \frac{1}{n}\sum_j \sum_r e_{jr}$$

assuming σ_{ej}^2 constant, the variance is

$$V(\bar{\bar{y}}) = \frac{\Sigma n_j^2}{n^2}\sigma_\gamma^2 + \frac{\sigma_e^2}{n} = \left(\frac{1}{k} + \frac{\Sigma(n_j - \bar{n})^2}{n^2}\right)\sigma_\gamma^2 + \frac{\sigma_e^2}{n} \qquad (4.5.3)$$

Unless the n_j vary greatly, the coefficient of σ_γ^2 is usually little larger than $1/k$.

Comparing $V(\bar{\bar{y}}_c)$ with $V(\bar{\bar{y}})$ from (4.5.2) and (4.5.3) respectively, we note that the coefficient of σ_γ^2, the between-cluster component of variance, is slightly smaller in $V(\bar{\bar{y}}_c)$, while that of σ_e^2 is slightly smaller in $V(\bar{\bar{y}})$ since $n > k\bar{n}_h$. The differences in variance are usually only moderate unless the n_j vary widely.

An unbiased sample estimate of $V(\bar{\bar{y}})$ can be constructed from an analysis of variance of the sample data. The expected values of the mean squares s_b^2 (between clusters) and s_w^2 (within clusters) work out as follows, where $Y_{j.}$ is a cluster total:

$$s_b^2 = \frac{1}{k-1}\left(\sum \frac{Y_{j.}^2}{n_j} - \frac{Y_{..}^2}{n}\right); \qquad E(s_b^2) = \sigma_e^2 + \bar{n}'\sigma_\gamma^2$$

and

$$s_w^2 = \frac{1}{n-k}\left(\sum \sum y_{jr}^2 - \sum \frac{Y_{j.}^2}{n_j}\right); \qquad E(s_w^2) = \sigma_e^2$$

where

$$\bar{n}' = \frac{1}{k-1}\left(n - \frac{\Sigma n_j^2}{n}\right)$$

(usually slightly less than $\bar{n}$). An unbiased sample estimate of $V(\bar{y})$ is obtained by inserting s_w^2 and $(s_b^2 - s_w^2)/\bar{n}'$ as estimates of σ_e^2 and σ_γ^2 in (4.5.3). This method would enable us to attach an estimated standard error to an estimate $\bar{y}$ from a completed survey.

In estimating k for a new survey from past data having similar clusters, one approach is to rewrite (4.5.3) in the form

$$V(\bar{y}) = \frac{1}{k}\left[\sigma_\gamma^2\left(1 + \frac{(k-1)(CV)^2}{k}\right) + \frac{\sigma_e^2}{\bar{n}}\right]$$

where $(CV)^2$ is the square of the coefficient of variation* of the cluster sizes n_j. The quantities σ_γ^2, σ_e^2, and $(CV)^2$ could all be estimated from past data and the relation between $V(\bar{y})$ and k estimated.

Persons unfamiliar with the implications of cluster sampling might use the estimate s^2/n for $V(\bar{y})$, where s^2 is the usual variance between individuals in the sample. It turns out that s^2/n is an underestimate, since

$$E\left(\frac{s^2}{n}\right) = \frac{\sigma_e^2}{n} + \frac{1}{n-1}\left(\frac{k-1}{k} - \frac{\Sigma(n_j - \bar{n})^2}{n^2}\right)\sigma_\gamma^2 \qquad (4.5.4)$$

By comparison with (4.5.3) the coefficient of σ_e^2 is correct, but that of σ_γ^2 is too small, being less than $1/(n-1)$ in $E(s^2/n)$ but greater than $1/k$ in the true variance in (4.5.3). The underestimation can be serious if the clusters are large (n/k large) or if members of a cluster give similar responses, so that the σ_γ^2 dominates σ_e^2.

To illustrate, suppose $k = 10$ clusters of sizes $n_j = 10, 12, 14, 16, 18, 22, 24, 26, 28, 30$, giving $n = 200$, $\bar{n} = 20$, and $\bar{n}_h = 17.62$. We find

$$V(\bar{y}) = 0.005\sigma_e^2 + 0.111\sigma_\gamma^2$$

$$V(\bar{y}_c) = 0.00567\sigma_e^2 + 0.1\sigma_\gamma^2$$

---

*The coefficient of variation is the standard deviation divided by the mean; here it would be the standard deviation of the cluster sizes divided by their mean size.

and

$$E\left(\frac{s^2}{n}\right) = 0.005\sigma_e^2 + 0.00447\sigma_\gamma^2$$

The coefficient of σ_γ^2 in $E(s^2/n)$ is only about $\frac{1}{25}$ of the correct value in $V(\bar{\bar{y}})$. This ratio is usually near $1/\bar{n}$ (in this example $\frac{1}{20}$).

Estimation of sample size naturally becomes more difficult in samples of complex structure, since we have to develop the correct variance formula and find a previous study of the same structure with the same response variable. In agencies that regularly employ complex survey plans, a helpful device used by some writers is the following [see Kish (1965), Section 8.2]. Keep a record of the ratio of the unbiased estimate of a quantity like $V(\bar{\bar{y}})$, computed from the results of the complex survey, to the elementary but biased estimate (in this case s^2/n) given by the standard formula that assumes a random sample of individual persons. This ratio is called the design effect (deff). For designs of the same complex structure, the deff ratios are often found to be closely similar for related response variables y, y', y'', and so forth. Consequently, if a previous sample of similar structure can be found when we are planning the size of a new sample, a knowledge of these deff ratios helps in determining a realistic estimate of sample size. Even if the y variable in the new survey was not measured in the previous survey, we may be able to guess a deff ratio for this y from the ratios for related variables in the previous survey. For example, suppose that the elementary formula for $V(\bar{\bar{y}})$ suggests $n = 500$ to make $V(\bar{\bar{y}}) = 2$. If we guess a deff ratio of around 1.3 for a sample of the intended structure, we increase n to $(500)(1.3) = 650$.

4.6 PLANS FOR REDUCING NONRESPONSE

The term "nonresponse" is used to describe the situation in which, for one reason or another, data are not obtained from a planned member of a sample. My impression is that standards with regard to nonresponse rates are lax in observational studies; one can name major studies in which nonresponse rates of 30–40% are stated with little reported evidence of earlier attempts to reduce these high figures.

As is well known, the primary problem created by nonresponse is not the consequent reduction in sample size; this could be compensated for by planning an initial sample size larger than needed. The real problem is that nonrespondents, if they could be persuaded to respond, might give somewhat different answers from the respondents, so that the mean of the sample of respondents is biased as an estimate of the population mean. We can think of a population as divided into two classes. Class 1, with mean $\bar{Y}_1$, consists of those who would respond to the planned sample approach. The mean $\bar{y}_1$ of the sample respondents is an unbiased estimate of $\bar{Y}_1$. Class 2, with mean $\bar{Y}_2$, consists of those who would not respond. If W_1 and W_2 are the population proportions of respondents and nonrespondents, the sample estimate $\bar{y}_1$ has bias $\bar{Y}_1 - \bar{Y} = \bar{Y}_1 - W_1\bar{Y}_1 - W_2\bar{Y}_2 = W_2(\bar{Y}_1 - \bar{Y}_2)$. With this simple model the bias does not depend on the sample size, so that the bias can dominate in large samples.

Sample evidence regarding whether $\bar{Y}_1$ and $\bar{Y}_2$ are likely to differ much is of course difficult to obtain, since this involves collecting information about those who were initially nonrespondents in a sample. Such information as has been collected indicates that (1) there is usually some nonresponse bias of size $\bar{Y}_1 - \bar{Y}_2$ depending on the type of question asked and on the sample approach, and (2) the bias is not necessarily serious, but it can be. From the form of the bias, $W_2(\bar{Y}_1 - \bar{Y}_2)$, the danger of any serious bias can be kept small by keeping W_2, the nonresponse rate, small.

Fortunately, W_2 can often be materially reduced by a combination of hard work and advance planning in anticipation of a nonresponse problem. The strategy adopted for reducing W_2 will depend on one's concept of the reasons for nonresponse in a planned sample.

Consider a survey in which the approach is directly to the individual member in the sample (either by mail, telephone, or household interview). A good attitude to keep in mind is that you are asking the sample member in effect to work for you (nearly always without pay) and that the member is busy. Usually the investigator opens with a brief account of the topic of the survey, stressing its importance and the reasons why the information is needed. It is helpful to capture the respondent's interest, but this depends on the topic. Additionally, the list of questions should be designed to convince the member that you are competent and are neither wasting the member's time, prying unnecessarily, nor asking the member to respond to vaguely worded questions.

Careful thought must be given to the order in which questions are to be asked. Early questions should be important and obviously relevant to the topic. For instance, as a professor I receive questionnaires about teaching practices and about attitudes or performance of the students. If the questionnaire begins with numerous questions about my past that do not seem relevant to what the investigator has stated he is trying to learn, the probability increases that I will be a nonrespondent. The same is true if there are questions such as "How would the students react if such and such a change were made?" My reply would be "Don't ask me, ask the students," or "Don't ask anybody—the students won't know either." Answers to hypothetical questions can seldom be trusted.

Every question that the investigator proposes beyond those obviously relevant, should be justified by considering: "Is this question essential?" The investigator should know the specific role that the answer to this question will play in the analysis, and how the analysis will be weakened if this question is omitted. The nonresponse rate usually increases as the questionnaire lengthens. If a batch of questions that are needed are likely to seem irrelevant to the respondent, it may be worth inserting a brief explanation as to why these questions are essential.

Devices that save the respondent's time should be sought, for example, indicating answers to questions by placing X's in boxes instead of writing answers. Using X's is feasible only when a limited number of types of answers to a question will cover the great majority of sample members, leaving an "other" written category for those who do not choose one of the boxes. With this scheme, some pilot checking is advisable to verify that the "other" category seems small, or to add one or two additional boxes.

Give an assurance of anonymity to each sample member. Except possibly in studies in which different questionnaires are sent to the same respondents at intervals of time, the respondent's name and address may not be needed on a returned questionnaire. If not, an identifying number on a mailed questionnaire will be necessary because you will want to know the names and addresses of those who did not respond to the first mailing in order that further mailings may be made to them.

In this connection, make advance plans for a definite *call-back* or *repeated-mailings* policy on those who do not answer the first inquiry. A minimum of up to three calls is considered advisable, while high-quality studies may insist on as many as six calls, if necessary. At the same time the effect of the call-back policy on the costs and the timing of the analysis and reporting of results needs to be considered. Actually, field results show that if the cost of planning the sample and the cost of conducting the statistical analysis are included, the overall cost per completed questionnaire is little higher for a three- or even a six-call policy than for a one-call policy, but time of analysis is affected. Comparison of results for the first and each later call provides clues about the nature of nonresponse bias.

In some studies, for example, of schools or branches of a business, the situation is that if the governing bodies of these establishments are convinced of the importance of the study, they will assure that the questionnaires are answered, except for reasons such as illness. Success or failure of a study may depend largely on the amount of planning, consultation, and discussion needed in presenting the case for the study before these governing bodies. The nonresponse problem here occurs in lumps; that is, refusal by a governing body may mean that 10 or 20% of the sample is missing in one decision. Persuading such a governing body to change its mind (the analogy of a successful call-back) challenges the ingenuity of the investigator.

4.7 RELATIONSHIP BETWEEN SAMPLED AND TARGET POPULATIONS

At some point in the planning it is well to summarize one's thinking about the relationship between the sampled populations—the populations from which our comparison groups will be drawn—and the target population to which we hope that the inferential conclusions will apply. Availability and convenience play a role, sometimes a determining role, in the selection of sampled populations. On reflection these may be found to differ in some respects from the target population. This issue is not confined to observational studies. For instance, controlled experiments in psychology may be confined to the graduate students in some department or to volunteer students at $5 per hour, although the investigator's aim is to learn something about the behavior of graduate students generally or even of all young people in this age range in the country. Airline pilots might be a convenient source of data for an inexpensive study of men's illnesses in the age range 40–50, but we hope that they are not typical of men, generally, in the frequency or severity of strokes or heart attacks.

The problem is that results found in the sampled populations may differ more or less from those that would be found in the target population. In studies of the economics of farming, a good source is a panel of farmers who regularly keep careful records of their economic transactions, in cooperation with a state university. But as would be expected, there is evidence (Hopkins, 1942) that such farmers receive a higher economic return from capital invested on their farms and adopt improved techniques more rapidly than do farmers generally.

This issue is common in program evaluation also, for example, teaching programs or client-service programs. Owing to difficulty of taking accurate research measurements within an operating program, it may be decided to conduct the study *outside* the program, although its results are intended to apply to the program. The change in the setting can affect the results. If the study is conducted *inside* the program, workers in the program, aware that they are being tested, may perform better in the study than they usually do in the ordinary operation of the program. Alternatively, the study, if imposed as a temporary extra load of work, might result in a lower quality of performance than is regularly attained. If a change in procedure in the program has been decided, the old and the new procedure may both be continued for a time in order to measure the size of the presumed benefit from the change. In this event, the workers, aware that the old procedure is to be abandoned, may do only slipshod work on it during the trial period, resulting in an overestimate of the benefit, if any, from the change.

The particular years in which the study is done may affect the results. A comparison of public versus slum housing might give one set of results in a period of full employment and rising prosperity, during which the slum families have the resources to move to superior private housing, but a different set if unemployment is steady and money is scarce. A well-planned series of experiments on the responses of sugar beet to fertilizers at the major centers in England was conducted at about 12 stations each year. After three years an argument arose for stopping the experiments because effects, while profitable, had been rather modest from year to year; the average responses to 90 lbs. nitrogen per acre were 78, 336, and 302 lbs. sugar per acre in the three years. There seemed little more to be learned. A decision to continue the experiments was made, however, because all three years had unusually dry summers. In the next two years, both wet years, the average effects of nitrogen rose to 862 and 582 lbs. sugar per acre.

Reflection on likely differences between sampled and target populations has occasionally caused investigators to abandon a proposed plan for a study, because the only available locale seemed so atypical of the target population that they doubted whether any conclusions would apply. Sometimes, the choice between two locales, otherwise about equally suitable, was made on this criterion. If resources permit, it might be decided to conduct the study in each of two locales that were atypical in different respects, or to have two or three control groups instead of only one, for instance where a new urban-renewal program is being tried in one town, and the "control" has to come from neighboring towns.

Supplementary analyses may help in speculating whether results obtained in one population are likely to hold up in another population. For instance, since different populations usually show somewhat different distributions of ages, economic levels, sex ratio, and urban–rural ratio, a statistical examination is relevant for this problem in a study that reveals the extent to which an estimated treatment effect varies with the level of any of these variables. The investigator might well regard it as part of his/her responsibility to report any aspect of the results or any feature of the sampled population that is similarly relevant. Research data on methods of handling some important social problems are scarce. An administrator in Washington,

D.C. or in California is likely to use the results of any study that can be found for policy guidance, for example, a study conducted on a particular group of people in Manhattan.

4.8 PILOT STUDIES AND PRETESTS

Early in the planning the investigator should begin to consider what can be learned from pilot studies and pretests. Most written discussions of the role of pilot studies deal with household-interview surveys (and to some extent with telephone or mail surveys), in which information can be gained about such issues as:

1. Ability of the interviewers to find the houses.
2. Adequacy of the questionnaire; for example, do some questions elicit many refusals, or many "don't knows" that could perhaps be avoided by a change in the form or the ordering of the questions.
3. Some indication of nonresponse rates.
4. A check on advance estimates of time per completed questionnaire per house and of costs of the field work.
5. Trial training for the interviewers.
6. If the planning team is undecided in selecting alternative forms of some questions, the effects of different question ordering, or household interviews versus telephone interviews with household interview used only in follow-up, a proposed pilot sample can be divided into random halves, using one alternative in each half.

One question is: Need the pilot sample be a random subsample of the whole planned sample, versus one chosen for speed and convenience in an area easily accessible to the planning headquarters? My opinion leans toward the latter choice, provided that the chosen pilot sample is judged to be reasonably representative of the range of field problems; for instance, we would obviously not want a pilot sample confined to rich person's houses if the questionnaire problems are likely to be relevant among the poor. If the pilot sample were intended to estimate variability for determination of sample size, it would have to be a random subsample, but pilot samples are used for this purpose only in planning major and expensive surveys in which substantial time and resources for a pilot study are considered essential. Parten (1950) and Moser (1959) provide good references to the roles of pilot samples in surveys.

If an observational study is to be conducted from existing records on individual persons, originally collected for another purpose, a pilot study of these records is highly advisable before committing oneself to the proposed study. Items to check include completeness, signs of gross errors, understanding of definitions, and signs of subtle changes in the meaning of the terms over time—or more generally, to provide an appraisal of the quality of the records for the intended purpose. It may be quite convenient to draw, say, an "every kth" systematic sample with a random start for the pilot. Time must be provided for the necessary statistical analysis of this pilot and for attempted follow-up of signs that arouse suspicion.

In a study to be done from state, regional, or national summary data, one goal of pilot work is to learn as much as possible about completeness of the data and any known or suspected biases. Useful strategies include discussions with persons involved in the collection of these data, searches for critiques of the data by outside persons, and preliminary graphical analyses (e.g., looking for sudden departures from smooth curves) followed by further discussion.

Since observational studies vary widely in nature, a further listing of possibilities will not be attempted. Try to plan any pilot work to aid a specific decision about the conduct of the study, rather than just having a look at how things go.

4.9 THE DEVIL'S ADVOCATE

When the plans for the study near completion, consider presenting a colleague with a fairly detailed account of the objectives of the study and your plans for it, and ask that person to play the role of devil's advocate by finding the major methodological weaknesses of your plan. It may be difficult to persuade this person to do this, since you are usually requesting more than a trifling amount of work. On the other hand, some scientists enjoy criticizing another's work and are good at it.

I stress this point because most observational studies, particularly those of any complexity, have methodological weaknesses. Some weaknesses are unavoidable due to the nature of an observational study or to the types of comparison groups available to us. Some weaknesses could be removed either by collecting data that we did not intend to collect at first or by using a more searching statistical analysis, as is seen when the investigator tries to reply to a slashing critique of his/her results that appears after the study has been published. Some readers may see the critique, but may see neither the investigator's original study nor the rebuttal. For some weaknesses that cannot be removed from the study, additional data or analyses may enable the investigator to reach a judgment regarding the strength of the objection, which the investigator can then publish as part of the report. Published results of studies on current social problems often rouse emotional reactions for or against the conclusions of the investigator. The reader who is emotionally against the investigator's conclusions is apt to magnify any criticism of a study that appears later. For this reason, a good practice in reporting is to list and discuss any methodological weakness or possible objection that has occurred to the investigator.

4.10 SUMMARY

Statistical theory provides formulas that aid in the estimation of the size of samples needed in a study. These formulas are worth using even when the choice of sample size is dominated by considerations of cost or number of available subjects. For exploratory studies, where the issue is whether a given agent or treatment produces any effect, the formulas estimate the sample size that will ensure a high probability of finding a statistically significant effect of the treatment when the real effect has a specified size δ. If the objective is estimation, the formulas give the size needed to make the estimate correct to $\pm L$ with high probability. Illustrations of the uses of the formulas in some simple problems are presented.

In observational studies the primary difficulties in using the formulas are (1) estimates of population parameters that appear in the formulas must be inserted, (2) biases increase the type-I errors in tests and decrease the probability that the estimated treatment effect is correct to within the stated limits, (3) the samples are often of more-complex structure than assumed in the simple formulas. An example of this type is given in which the sample consists of groups or clusters of subjects, rather than individual subjects.

The term "nonresponse" is given to the failure to obtain some of the planned measurements. The problem with nonresponse is not so much the reduction in sample size, which can be compensated for by a planned sample larger than is needed. There is, however, evidence that people unavailable or unwilling to respond may differ systematically from those who respond readily, so that results from the respondents are biased if applied to the whole population. In planning, likely sources of nonresponse need to be anticipated and plans need to be made to keep the level of nonresponse low. Repeated call-backs are a standard device in mail and household-interview surveys. Questionnaires should be constructed so as to gain the respondent's interest, respect, and confidence. Sometimes, the main hurdle is to devise an approach that will obtain permission and support from an administrative or governing body.

In both observational studies and controlled experiments, the population represented by the study samples may differ from the target population to which the investigator would like the results to apply. At some point in the planning, the investigator should reflect on the differences between the sampled and target populations; sometimes, supplementary analyses can be carried out that help in judging to what extent results for the sampled population will apply to the target population.

The reasons for a pilot study on some aspect of the proposed plan should be considered. In an interview survey, pilot studies can gain information on such matters as wording, understanding and acceptability of the questions, the sources and nature of the nonresponse problem, the time taken, and field costs. In a study of existing records, completeness and usability of the records for research purposes can be checked, and, more generally, any uncertain aspect of the proposed plan.

When the proposed plan nears completion, a colleague capable of critiquing the plan can help by reviewing the plan, pointing out methodological weaknesses that have escaped the notice of the planners, and, if possible, suggesting means of remedying these weaknesses. The report of the results should discuss weaknesses that cannot be removed from the plan, and give the investigator's judgment regarding the effects of the weaknesses on the results.

REFERENCES

Hopkins, J. A. (1942). Statistical comparisons of record-keeping farms and a random sample of Iowa farms for 1939." *Agr. Exp. Sta. Res. Bull.*, **308**, Iowa State College.

Kish, L. (1965). *Survey Sampling*. Wiley, New York.

Moser, C. A. (1959). *Survey Methods in Social Investigation*. Heinemann, London.

Parten, M. B. (1950). *Surveys, Polls and Samples: Practical Procedures*. Harper & Brothers, New York.

CHAPTER 5

Matching

5.1 CONFOUNDING VARIABLES

When we plan to compare the mean responses y in two or more groups of subjects from populations exposed to different experiences or treatments, the distributions of y, ideally, should be the same in all populations, except for any effects produced by the treatments. In fact, the values of y are usually influenced by numerous other variables $x_1, x_2, \ldots$ which will be called *confounding variables*. They may be qualitative or ordered classifications, discrete or continuous.

Confounding variables may have two effects on the comparison of the response y in the two populations. First, since in an observational study the investigator has limited control over his choice of populations to be studied, the distributions of one or more of the confounding variables may differ systematically from population to population. As a result, the distributions of y may also differ systematically and the comparison of the sample mean values of y may be biased. Second, even if there is no danger of bias—the distributions of a confounding variable x being the same in different populations—variations in x contribute to variability in y and decrease the precision of comparisons of the sample means.

Two simple examples will be given to show how a confounding variable may produce bias and decrease the precision of a comparison $\bar{y}_1 - \bar{y}_2$. The examples also suggest the two principal techniques used in practice to control undesirable effects of a confounding variable.

In the first example there are two samples, treated (t) and control (c), and y has a linear regression on a confounding variable x, of the same form $\alpha + \beta x$ in each population. Hence

$$y_{ti} = \alpha + \delta + \beta x_{ti} + e_{ti}$$

and

$$y_{ci} = \alpha + \beta x_{ci} + e_{ci}$$

where δ is the effect of the treatment. As given previously, the sample mean difference $\bar{d}$ is

$$\bar{d} = \bar{y}_t - \bar{y}_c = \delta + \beta(\bar{x}_t - \bar{x}_c) + (\bar{e}_t - \bar{e}_c) \qquad (5.1.1)$$

and

$$E(\bar{d}) = E(\bar{y}_t - \bar{y}_c) = \delta + \beta(\mu_{tx} - \mu_{cx}) \qquad (5.1.2)$$

so that the bias is of amount $\beta(\mu_{tx} - \mu_{cx})$. Further, regardless of whether $\mu_{tx} = \mu_{cx}$, we have from (5.1.1), assuming the variances of x and e are the same in each population,

$$V(\bar{d}) = V(\bar{y}_t - \bar{y}_c) = \frac{2}{n}\left(\beta^2 \sigma_x^2 + \sigma_e^2\right) = \frac{2}{n}\left(\rho^2 \sigma_y^2 + (1 - \rho^2)\sigma_y^2\right)$$

It follows, as is well known, that if we could remove the effects of variations in x, we could reduce $V(\bar{d})$ from $2\sigma_y^2/n$ to $2(1 - \rho^2)\sigma_y^2/n$. This is the main reason for the use of the methods known as the *analysis of covariance* and *blocking* in controlled experiments. By the random assignment of subjects to treatment groups and by other precautions, the investigator in a simple controlled experiment hopes that he does not have to worry about bias in the comparison $\bar{d}$. It may still be worth trying to control confounding x variables for the potential increase in the precision of $\bar{d}$.

If the linear-regression model is correct, Eq. (5.1.1) suggests two alternative methods of removing the danger of bias and increasing the precision. The first, used at the planning stage, is to select the treatment and control samples so that $\bar{x}_t$ and $\bar{x}_c$ are equal, or nearly equal. In repeated samples of this type,

$$\bar{d} \doteq \delta + (\bar{e}_t - \bar{e}_c) \qquad (5.1.3)$$

giving $E(\bar{d}) \doteq \delta$ and $V(\bar{d}) \doteq 2\sigma_e^2/n = 2\sigma_y^2(1 - \rho^2)/n$
The second method is applied at the analysis stage. From (5.1.1),

$$\bar{d} = \delta + \beta(\bar{x}_t - \bar{x}_c) + (\bar{e}_t - \bar{e}_c)$$

Hence, we compute an estimate $\hat{\beta}$ of β and estimate δ by the adjusted mean difference

$$\bar{d}' = \bar{d} - \hat{\beta}(\bar{x}_t - \bar{x}_c) = \delta + (\bar{e}_t - \bar{e}_c) - (\hat{\beta} - \beta)(\bar{x}_t - \bar{x}_c)$$

If $\hat{\beta}$ is an unbiased estimate of β, then $E(\bar{d}') = \delta$. If the effects of sampling errors in $\hat{\beta}$ were negligible, we would have $V(\bar{d}') = 2\sigma_y^2(1 - \rho^2)/n$. These sampling errors increase this value, but only trivially in large samples.

To summarize for this example, we should plan to control an x variable either if there seems to be a danger of nonnegligible bias or if a substantial gain in precision may result. Using a linear regression, we do not begin to get a substantial reduction in $V(\bar{d})$ until ρ is at least 0.4. An x variable can be controlled either by the way in which the samples are selected in the planning stage or by recording the values of x and adjusting the estimate in the analysis stage.

As a second example, suppose that x is a two-class variate and y is a proportion calculated from a $0 - 1$ variate. For the treatment (t) and control (c) populations, the proportions and the means of y in each class are given in the list below.

Sample	Population Proportion in x: Class 1	Class 2	Population Mean of y x: Class 1	Class 2	x: Overall
Treatment	f_t	$(1 - f_t)$	$\delta + p_1$	$\delta + p_2$	$\delta + f_t p_1 + (1 - f_t)p_2$
Control	f_c	$(1 - f_c)$	p_1	p_2	$f_c p_1 + (1 - f_c)p_2$

The population means of y differ in the two classes, having values p_1 and p_2 for the control population. The true treatment effect is δ in each class. Each sample has total size n, but the expected proportions f_t and f_c that fall in class 1 have been made to differ for the treatment and control samples. This difference is the source of the trouble.

If the overall treatment and control samples are randomly drawn from the populations composed of classes 1 and 2, the sample proportions $\hat{p}_t$ and $\hat{p}_c$ have means as shown in the right-most column of the list. It follows from the list that

$$E(\bar{d}) = E(\hat{p}_t - \hat{p}_c) = \delta + (f_t - f_c)(p_1 - p_2)$$

Thus there is a bias of amount $(f_t - f_c)(p_1 - p_2)$. Note that a large bias requires both a large difference in the expected proportions f_t and f_c in class 1 and a large difference in the means p_1 and p_2 in the two classes. This explains why there is sometimes only a small bias in the estimated difference $\hat{p}_t - \hat{p}_c$, even if f_t and f_c differ widely.

This result also suggests two methods of controlling bias, analogous to the methods given in the first example. At the planning stage, we could draw samples subject to the restriction that $f_t = f_c$, but otherwise drawn at random. This technique is frequently referred to as "within-class matching" or "frequency matching." Alternatively, without this restriction, we could adopt a different estimate of δ at the analysis stage. Let $\hat{p}_{t1}$, $\hat{p}_{t2}$, $\hat{p}_{c1}$ and $\hat{p}_{c2}$ be the sample estimates of the treatment and control proportions in each class from random samples. Any weighted estimate of the form

$$\bar{d}' = W_1(p_{t1} - p_{c1}) + W_2(p_{t2} - p_{c2}) \qquad (W_1 + W_2 = 1)$$

is clearly an unbiased estimate of δ under this model. In practice, the particular choice of W_1 and W_2 has varied a good deal from study to study. Some investigators choose the weights W_1 and W_2 from a standard population that is of interest (perhaps the target population), and others choose W_1 and W_2 to minimize the variance of $\bar{d}'$.

In this example there is no danger of bias if either $p_1 = p_2$ or $f_t = f_c$. Do the methods of controlling bias increase the precision in the "no bias" situations, as they did in example 1? There is no increased precision if $p_1 = p_2$. If $p_1 \neq p_2$ there is a possible increase in precision if we make $f_t = f_c$, but this is small unless p_1 and p_2 differ greatly. In my opinion this is seldom worth any extra trouble in practice.

To summarize for this example, attempts to remove bias or increase precision may be made either in the planning or the analysis stages. However, when y is a proportion and there is no danger of bias, the increase in precision resulting from these attempts is usually small or moderate.

In handling the problem of confounding variables, the investigator should first list the principal confounding variables that he recognizes, in order of their importance in influencing y, inasmuch as this can be judged. A decision is made to exercise some control over an x variable either if the possibility of a nonnegligible bias exists in the y comparison or if a substantial gain in precision may result.

A second requirement about any x variable that we plan to control is that its value should not be influenced by the treatments to be compared. Suppose that in the first example the values of x are higher in the treatment than in the control population because the treatment affects x, and that x and y are positively correlated. In this case, subtracting $\hat{\beta}(\bar{x}_t - \bar{x}_c)$ from $\bar{d}$ removes part of the treatment effect on y.

This mistake is avoided when the x variables are measured before the introduction of the treatment, but the danger exists whenever the measurement is subsequent to the introduction of the treatment. For example, Stanley (1966) cites a study intended to measure the effect of brain damage existing at birth on the arithmetic-reasoning ability of 12-year-old boys. In comparing samples of brain-damaged and undamaged boys, the investigator might be inclined to use current measures of other kinds of ability, or parental socioeconomic status as confounding x variables whose effects are to be removed by matching or regression adjustment. Other kinds of ability at age 12 might obviously be affected by the brain damage, and as Stanley points out, even parental socioeconomic status might also be affected because of the strain and cost of medical care for the brain-damaged child.

In the subsequent discussion of specific techniques for handling confounding variables, it is necessary to keep in mind the scales of measurement of both x and y. With x, the principal distinction is between a classification and a discrete or continuous variable; with y the principal distinction is between a proportion derived from a $0 - 1$ variate and a continuous or discrete variable. This chapter discusses methods used at the planning stage for handling confounding variables. Adjustments in analysis are discussed in Chapter 6.

5.2 MATCHING

In matching, a confounding x variable is handled at the planning stage by the way in which the samples for different treatment groups are constructed. In some methods each member of a given treatment group has a match or partner in every other treatment group, where the partners are within defined limits in the values of all x variables included in the match. The number of x variables matched in applications may range from 1 to as many as 10 or 12.

From inspection of medical journals, Billewicz (1965) reports that the numbers of variables most often matched in medical studies were two or three. The idea of "matching" is the same as that known as "pairing" or "blocking" in experimentation. As a rule, matching is confined to smaller studies of simple structure—most commonly, two-group comparisons. The more complex the plan, the more difficult it will be to find matches. What is meant by a match? This depends on the nature of the confounding variable x.

x a Classification.   A match usually means belonging to the same class. With three classified x variables having two, four, and five classes, respectively, a match on all three variables is another subject in the same cell of the $2 \times 4 \times 5 = 40$ cells created by this three-way classification.

x Discrete or Continuous.   Two procedures are common: One is to change x into a classification variable (e.g., ages arranged in five-year classes) and as before regard a match as someone in the same class. This method is common if, say, two of three x variables are already classification variables, the third variable being originally continuous. This method will be called *within-class matching* (other terms used are "stratified matching" and "frequency matching").

With x discrete or continuous a second method is to call two values of x a match if their difference lies between defined limits $\pm a$. This method is called *caliper matching*, a name suggested by Donald Rubin (1970). With x continuous, within-class matching is more common than caliper matching, for which it is harder to find matches.

Two advantages of matching are that the idea is easy to grasp and the statistical analysis is simple. Perfect caliper matching on x removes any effects of an x variable, whatever the mathematical nature of the relation between y and x, provided that this relation is the same in the populations being compared. No assumption of a linear regression of y on x is required. To verify this, suppose that for the jth subject in population 1 the relation between y and x is of the general form

$$y_{1j} = \delta_1 + f(x_{1j}) + e_{1j}$$

In population 2,

$$y_{2j} = \delta_2 + f(x_{2j}) + e_{2j}$$

where $f(x)$ has any functional form and e_{1j} and e_{2j} have means zero. Then if $x_{1j} = x_{2j}$ for all j,

$$\bar{d} = \bar{y}_1 - \bar{y}_2 = \delta_1 - \delta_2 + \bar{e}_1 - \bar{e}_2$$

and is unbiased whatever the nature of the function $f(x)$. If matching is not perfect but fairly tight, the hope is that for any continuous function $f(x)$ we will have $f(x_{1j}) \doteq f(x_{2j})$ because $x_{1j} \doteq x_{2j}$, and the remaining bias in $\bar{y}_1 - \bar{y}_2$ will be small.

Matching has some disadvantages. The long time taken to form matches, may hardly seem worthwhile if under the original matching rules no matches can be found for some members of one sample. Imperfect matching on the chosen variables or omission of important variables on which we failed to match can leave systematic differences between the members of a matched pair. Billewicz cites an example by Douglas (1960), in which children from premature births ($5\frac{1}{2}$ lb or less) were found to have inferior school performance at ages 8 and 11 to normal-birth children. The original sample size was 675; samples were matched with regard to sex, mother's age, social class, birth rank in the family, and degree of crowding in the home. It became evident, however, as the study progressed that despite the matching, systematic differences remained between the parents of prema- ture- and normal-birth children regarding (1) social level, (2) maternal care, and (3) interest in school progress. Each matched pair of children was assigned a score (from $+3$ to -3) according to the extent to which these three variables favored the premature child. The mean differences in the exam results (of 11-year-old premature- and normal-birth children) ap- peared as follows when subclassified by this score. (Results for the exam of eight-year-olds were similar.)

Premature–Normal Scores On Confounding Variables	Exam Scores
$+3$	$+6.0$
$+2$	$+0.4$
$+1$	-0.6
0	-1.7
-1	-5.6
-2	-6.7
-3	-12.0

Clearly, the original matching did not guarantee that partners were equiva- lent on all important confounding variables, even after matching on five variables. (Firm interpretation of this finding rests, in part, on being certain that neither maternal care nor interest in school progress is affected by the comparison variables—premature versus normal birth.)

With x continuous, some results for two other matching methods will be presented later in this chaper. One method called *mean matching* (or "balancing") does not attempt to produce closely matched individual pairs, but instead concentrates on making $\bar{x}_1 - \bar{x}_2$ as small as possible. This method is not new. It is of course tied to the assumption that y has the same linear regression on x in each population. The second method, called *nearest available matching*, tries to produce well-matched pairs in difficult situations and is described later.

In studying the performance and properties of various matching proce- dures, we shall consistently use x_1, n_1, σ_1, and so forth, to relate to the group of observations *for* which matching observations are being sought. They are sought *from* reservoir 2, characterized by entities x_2, n_2, σ_2, and so forth, all bearing the subscript "2." This notational asymmetry is critical for correctly interpreting tables and formulas.

5.3 THE CONSTRUCTION OF MATCHES

Published reports of studies using matching illustrate that the practical difficulties in constructing matched samples vary greatly from study to study. Numerous factors are relevant. In order to form a matched sample of size n from two or more populations, we obviously need larger supplies or reservoirs of subjects from which matches may be sought. The sizes and accessibility of these reservoirs are important. The most difficult case is one in which the investigator has only n subjects available from one population and needs all of them. Unless the reservoir from a second population is much larger than n, the investigator may be unable to find matches for all n from population 1.

Another factor is the planned size of the sample to be compared. Matching is seldom used when this planned n exceeds say 500, presumably because of the labor and time needed to match. The difficulties of finding matches under fixed rules also mount rapidly with each increase in the number of x variables to be matched. In Section 5.2 the Douglas example included a reservoir of 12,000 normal births, where $n = 675$ with five matched variables.

Matching of samples from two populations becomes more difficult when the x distributions differ markedly in the two populations—the situation in which the risk of bias due to x is greatest. If most people in population 1 are older than those in population 2, it may be impossible to find good matches for the oldest members of a sample from population 1. This difficulty is sometimes handled by omitting sample members who cannot be matched by the original rules, rather than by relaxing the matching rules. The full consequences of the possible alternatives have not been investigated. If the regression of y on x is the same function in both populations, omission may be the better procedure, though it means that one of the samples is badly distorted at one end.

The time taken to create matched samples depends much on the ease with which we can locate sample members whose x values are in the desired range. Sometimes, the available records are kept in a form that facilitates this search, as might happen if a match for a newborn baby is as follows: a child of the same sex, born in the same hospital during the same week with no complications of delivery. When one has to seek matches by going through a reservoir case by case, chance plays a major role in determining how long the job takes. For illustration, suppose that x is a five-class variable and that the sample from population 1 has only $n = 100$ cases available, distributed as follows:

Class	1	2	3	4	5	Total
Number	10	25	30	25	10	100

In population 2 the proportions falling in these five classes are assumed as follows:

Class	1	2	3	4	5	Total
Proportions	0.038	0.149	0.269	0.326	0.218	1.000

In classes 4 and 5, it appears that a search of less than 100 cases from population 2 should provide the needed 25 and 10 matches, but more than 100 seem necessary, on the average, for the other three classes. The most difficult case is class 1, where we need 10 matches but expect only 3.8 from a sample of 100. The number of reservoir cases needed to find these 10 matches is a random variable following a simple waiting-time distribution [Feller (1957)]. The mean number needed is $10/0.038 = 263$, or more generally m/p, where m is the number required and p is the proportion. This number has a large variance $m(1 - p)/p^2 = 3835$ in this example. The consequence is that the upper 95% point of the waiting-time distribution exceeds 350. Thus we might be lucky and find the 10 matches needed in class 1 in 150 cases, or we might be unlucky and have to search over 350 cases. This uncertainty makes the case-by-case construction of matches from random samples frustrating, particularly when there are several x variables.

For this reason, matching is inadvisable if potential sample members from the different populations become available at the rate of only a few per week, for example, subjects entering an agency for a service of some kind. There may be an indefinite delay while waiting for matches for certain subjects. This point is discussed more fully by Billewicz (1965).

Computers should be able to perform much of the detailed labor of finding matches if the values of the x variables in the available reservoirs are in a form suitable for input into computers. The easiest case is within-class matching when all the x variables, three for example, are already in classified form. For each reservoir, simple instructions will arrange and list in a printout the sample in each cell of the three-way classification. We learn, for instance, that the first reservoir has 19 cases in cell 2; the second reservoir has 28 cases. For any desired sample size up to 19 from this cell, the partners can be drawn at random by the computer from the 19 or 28 cases available. If two or three x variables to be matched are already in classified form, while the third is a continuous variable that is to be classified for within-class matching, the computer will perform this classification, arrange each reservoir in cells, and list as before.

With a single discrete or continuous x for which caliper matching is desired, the computer can rank and list the values of x in each reservoir from lowest to highest. From these lists caliper matches can be made quickly if available. I do not know how best to extend this method to two or three continuous x's where a caliper match is needed for each. A partial help is to have the computer classify each x into one of 2^m classes by binary splitting. Thus with three x's and four classes per variable, the computer arranges the trinomial distribution into 64 cells and lists. Since the values of x are not strictly in rank order within a cell, such lists are less convenient, but still a considerable help in searching for caliper matches on all three x's.

In matching samples from two populations with one continuous x, the investigator sometimes needs to use *all* cases in the reservoir from population 1. A sample of at least 100 cases is needed and there are only 100 cases from population 1. The reservoir from population 2 has say 272 cases. In this case it is not clear that caliper matching of each case in sample 1 can be performed with a prechosen fixed $\pm a$ value. The investigator does not want to reject any cases, since this reduces the sample size below 100. For this problem, Donald Rubin (1970) has developed a method called *nearest available pair matching*, performed entirely by computer, that attempts to do the best job of matching subjects to the restriction that every member of sample 1 must be matched.

The computer first arranges sample 1 in random order. For the first member of sample 1 it picks out the member of the reservoir for sample 2 that is nearest to it and lays this pair aside as the first match. The process is repeated for the second member of sample 1 with respect to the 271 items remaining in the reservoir, and so forth. Thus all matches are found, though, of course, the difference $|x_{1j} - x_{2j}|$ will differ from pair to pair.

Two variants of this method that might be better were also examined by Rubin. Instead of arranging sample 1 in random order the computer first ranks the sample 1 members from lowest x_1 to highest x_n. In variant 1 we seek matches from reservoir 2 in the order $x_n, x_{n-1}, \ldots$ (high–low). In variant 2 the order is $x_1, x_2, \ldots$ (low–high).

In mean matching, the objective is to make $\bar{x}_1 - \bar{x}_2$ as small as possible for any x. If all members of sample 1 must be used, their mean $\bar{x}_1$ is first found. The computer selects from reservoir 2 the value x_{21} nearest to $\bar{x}_1$. Then x_{22} is chosen such that $(x_{21} + x_{22})/2$ is nearest to $\bar{x}_1$ and so forth.

5.4 EFFECT OF WITHIN-CLASS MATCHING ON x

The problem of the effects of matching is complex, and not nearly enough is known about it. As we have mentioned, the purposes in matching are (1) to protect against bias in $\bar{y}_1 - \bar{y}_2$ that might arise from differences between the x distributions in different populations to be compared and (2) to increase the precision of the comparison of the y means. This section discusses the effect of within-class matching on x. We must first consider the nature of the x variable. There are three possibilities:

1. *An "Ideal" Classification.* This term is used for classifications in which two members of the same class are identical with regard to x. Within-class matching therefore gives perfect matching, which as previously noted removes bias in y for any functional form of the relationship between y and x that is the same in both populations. Unfortunately, it is not clear how often ideal classifications occur in practice. This might be so with a qualitative classification like the O, A, B, AB blood types in which two subjects with the same blood type are identical with regard to any effect of the relevant genes on y.

Sex (male, female) might be an ideal classification for some types of response y, but not for other types. For instance, in traits related to behavior or attitudes, it is natural to think of some women as more feminine than other women, and some men as more masculine than other men; then the male–female classification would have classes of nonidentical members.

2. *A Classification with Any Underlying Distribution.* Numerous examples can be given of classifications in which members of the same class need not be identical with regard to the variate which x is designed to measure. Consider an urban–rural classification. Many aspects of urban–rural living that are likely to influence y, are themselves affected by the fact that some people in the urban class have a more typically city environment than others in the urban class; likewise, some people in the rural class have a more typically country environment then others in the rural class. The same is true, for some responses, of a classification by religion into Catholic, Jewish, and Protestant. Some subjects of a given religion are much more heavily committed to religious beliefs and activities than are other subjects.

Ordered classifications such as socioeconomic level or degree of interest (none, little, much) in some topic are a more-obvious example. One can sense an underlying continuous x variable that has been divided into a small number of ordered classes. Indeed, ordered classifications are often used when we recognize that a correctly measured x would be continuous, but can measure only crudely, so that an ordered classification seems all that the measuring instrument will justify.

In examining the effect of within-class matching on bias when there is an underlying distribution, I assume that the correct x (the value that influences y) is continuous and that the observed classified x represents a grouping of the correct x into ordered classes. Consider how matching affects $\mu_1 - \mu_2$, the mean of $\bar{x}_1 - \bar{x}_2$. If $\mu_1 > \mu_2$ it seems plausible that within most classes the mean μ_{1j} will exceed μ_{2j}, where j stands for the class. This is true for common unimodal distributions such as two normals or two t variates with different means. In the matched samples the mean of $\bar{x}_1 - \bar{x}_2$ is $\Sigma W_j(\mu_{1j} - \mu_{2j})$, where W_j are the relative numbers in the classes. Thus it seems likely that even after matching, the mean of $\bar{x}_1 - \bar{x}_2$ will tend to be positive, though its size should be limited by the width of the classes.

3. *x Discrete or Continuous.* Since we are discussing within-class matching, we assume that discrete or continuous x's are grouped into a limited number of classes before matching. Consequently, mathematical study of the effect of matching on $\bar{x}_1 - \bar{x}_2$ follows the method just indicated, except that we are now interested also in the optimum choice of class boundaries and in the effects of different numbers of classes, since these are under our control when we create the classified x variable.

With x distributed as $N(B, 1)$ in population 1 and as $N(0, 1)$ in population 2, the percent bias removed by matching was first calculated for $0 \leqslant B \leqslant 1$ for specified division points $x_0, x_1, \ldots, x_c$ (with c classes). The value $B = 1$ was considered a larger initial bias than would be typical in practice. The percent bias removed is

$$100\left[1 - (\mu_1 - \mu_2)_m/B\right]$$

where $(\mu_1 - \mu_2)_m$ is the mean of $\bar{x}_1 - \bar{x}_2$ from the matched samples. The percent bias removed was found to be practically constant in the range $0 \leqslant B \leqslant 1$ and therefore could be approximated by calculus methods for B small [Cochran (1968)].

Let the distribution of x be $f(x)$ in population 1 and $f(x - B)$ in population 2, where $f(x)$ has unit SD (standard deviation). By the calculus method, the percent reduction in bias was found to be

$$100 \sum_{j=1}^{c} M_j\left[f(x_{j-1}) - f(x_j)\right] \tag{5.4.1}$$

where M_j is the mean value of x from $f(x)$ in the interval (x_{j-1}, x_j). In particular, for x normal,

$$M_j P_j = \frac{1}{\sqrt{2\pi}} \int_{x_{j-1}}^{x_j} x e^{-x^2/2}\, dx = f(x_{j-1}) - f(x_j)$$

where P_j is the total frequency in class j. Thus, for the normal, (5.4.1) gives

$$\text{Percent reduction in bias} = 100 \sum_{j=1}^{c} \frac{\left[f(x_{j-1}) - f(x_j)\right]^2}{P_j} \quad (5.4.2)$$

It happens that for the normal distribution, (5.4.2) is also the percent reduction in the *variance* of $\bar{x}_1 - \bar{x}_2$ due to matching, a result that follows immediately from results given in other cases by Ogawa (1951) and D. R. Cox (1957). For numbers of classes between 2 and 10, these authors also determined the optimum boundaries (shown here as the optimum relative class sizes), the corresponding maximum percent reductions in bias and variance for x normal, and the percent reductions when the classes are made equal in relative frequency. These values are given in Table 5.4.1.

With the optimum boundaries at least five classes are necessary to remove 90% or more of an initial bias in $\bar{x}_1 - \bar{x}_2$. Only 64% is removed with two classes and 81% with three classes. This is disappointing because matching with three classes is not uncommon, sometimes because only three classes are given in an ordered classification. With the optimum boundaries the central classes are larger, in terms of frequency, than the extreme classes. It is noteworthy, however, that with equal-sized classes the percentage reductions in bias and variance are only around 2% less than the maximum reductions. The choice of class boundaries and resultant sizes is not critical.

For equal-sized classes, some investigations of nonnormal distributions (Cochran, 1968) found that the percent reductions in bias agreed quite well

Table 5.4.1. Optimum Sizes of Classes and Percent Reductions in Bias and Variance of $\bar{x}_1 - \bar{x}_2$ Due to Within-Class Matching (x Normal)

Number of Classes	Optimum Class Frequencies (%)[a]	Percent Reductions	
		Maximum	Equal Classes
2	50	63.7	63.7
3	27, (46)	81.0	79.3
4	16, 34	88.2	86.1
5	11, 24, (30)	92.0	89.7
6	7, 18, 25	94.2	91.9
7	5.5, 14, 20, (21)	95.6	93.4
8	4, 11, 16, 19	96.7	94.5
9	3, 8, 13, 17, (18)	97.2	95.4
10	2, 7, 11, 14, 16	97.6	95.9

[a]Since the distribution is symmetrical, only the lower half is shown, starting with the lowest class. Thus for $c = 4$, the frequencies are 16, 34, 34, 16 in percentages; for $c = 5$, they are 11, 24, 30, 24, 11.

with those for the normal—running about 2% higher in some cases. In variance, the percent reductions tended to fall below those for the normal as skewness and kurtosis increase.

Unequal variances were also examined where x follows $N(B,1)$ in population 1 and $N(0, \sigma_2^2)$ in population 2. For values of σ_2^2 between $\frac{1}{2}$ and 2, it appeared that the percent reductions in the bias of $\bar{x}_1 - \bar{x}_2$ differed little from the values given for $\sigma_2^2 = 1$.

The effect of within-class matching on the bias of $\bar{x}_1 - \bar{x}_2$ are presented in Table 5.4.1. Results for the effects of caliper matching, nearest-neighbor matching, and mean matching on $\bar{x}_1 - \bar{x}_2$ will be given in Sections 5.5–5.7; the situation with respect to the bias of $(\bar{y}_1 - \bar{y}_2)$ will be discussed in Section 5.8.

5.5   EFFECT OF CALIPER MATCHING ON x

Caliper matching is a tighter and more-efficient method than within-class matching and can be used when x is continuous or discrete. For x continuous, there is a certain inconsistency in within-class matching. For example, when we search for a match for a subject whose true x value is 59.4, we may reject a subject whose x value is 60.2 because this subject is in the next-higher class, but we may accept a subject whose value is 42.1 because this subject is in the same class.

With caliper matching to within $\pm a$, the frequency functions of x in the two groups (populations 1 and 2) were assumed to be $N(B,1)$ and $N(0,1)$. As with within-class matching, the percent of the bias removed in $\bar{x}_1 - \bar{x}_2$ is fairly constant for values of an initial bias B which are typical of those values that occur in practice. For a given $f(x)$, the percent depends primarily on a or, more generally, on the ratio a/σ.

The amount of bias removed also depends, to some extent, on how the caliper matching is done. One method starts with a sample from population 1 and finds matches for all its members. Thus in the matched pairs, the members from sample 1 still represent an undistorted sample from population 1. However, if say $\mu_1 > \mu_2$, the matches selected will make the members from population 2 a selected sample that is biased upwards. If instead we start with ample reservoirs from both populations and search for the matches that can be found most quickly, we will tend to select members on the low-bias side from population 1 and on the high-bias side from population 2, since these are the easiest to match. An extreme form of this approach is to assume that we start with a random sample of the differences $x_{1j} - x_{2j}$ and go through this sample selecting the pairs that are caliper-matched. This approach results in smaller values of $|x_{1j} - x_{2j}|$ for matched pairs because of the distortion of *both* initial samples.

Table 5.5.1. Percent Reduction in Bias of $\bar{x}_1 - \bar{x}_2$
with Caliper Matching to Within $\pm a$ (Normal
Distribution). In (1) one Sample is Random and in
(2) Matches are Made from $x_{1j} - x_{2j}$

$\pm a$	Percent Reduction (1)	Percent Reduction (2)
0.2	99	99
0.3	97	98
0.4	95	97
0.5	93	96
0.6	90	94
0.7	87	92
0.8	84	90
0.9	80	87
1.0	76	84

For x normal, Table 5.5.1 shows the percent bias removed when (1) one
sample is random and (2) matches are made from random differences.
These are intended to indicate the range of performance in applications.

By comparison with Table 5.4.1, which gives percent reductions in bias
due to within-class matching, caliper matching to within $\pm 0.9\sigma$, which
seems quite loose, should be as good as within-class matching with three
classes, and removes 80% or more of the bias. Caliper matching to within
$\pm 0.4\sigma$ is as good as within-class matching with nine classes, and removes
95% of the bias. However, the benefits of caliper matching have a
cost—caliper matching, in general, requires much larger reservoirs and more
time.

When there is no bias, the percent reductions in the variance of $\bar{x}_1 - \bar{x}_2$
with caliper matching approximate the higher values for bias removed, that
is, the percent (2) values presented in Table 5.5.1.

With unequal variances in the two populations and x having frequency
functions $N(B, 1)$ in population 1 and $N(0, \sigma_2^2)$ in population 2, caliper
matching for a given a does a little better than the values presented in Table
5.5.1 when $\sigma_2^2 > 1$, and somewhat worse when $\sigma_2^2 < 1$.

5.6 EFFECT OF "NEAREST AVAILABLE" MATCHING ON x

Rubin's (1970) results for the effects on bias will be quoted for x normal.
[For a more extensive treatment see Rubin (1973, a, b).] It is assumed that
all n_1 available subjects from population 1 are to be matched, and that the

reservoir from population 2 has N_2 subjects. The effect of the technique naturally depends on the ratio N_2/n_1. It also depends on the size of the initial bias B, unlike the previous methods, because with B positive ($\mu_1 > \mu_2$) and substantial, we can expect only "poor" nearest neighbors for the highest members of sample 1.

Table 5.6.1 gives the percent reductions in bias of $\bar{x}_1 - \bar{x}_2$ for $n_1 = 50$, 100; $N_2/n_1 = 2, 3, 4$; and $B/\sigma = \frac{1}{4}, \frac{1}{2}, \frac{3}{4}, 1$, with σ assumed the same in both populations. My opinion is that $B/\sigma = \frac{1}{4}, \frac{1}{2}$ is more representative of the sizes of initial bias that occur in practice than is $B/\sigma = \frac{3}{4}, 1$, which seems unusually large, although initial age biases for cigar and pipe smokers in the studies on smoking (Section 2.4) were around 0.7σ.

The results suggest that with a reservoir in population 2 four times as large as the sample from population 1, the method should remove nearly all the bias for initial biases up to $\frac{3}{4}\sigma$. For moderate biases (up to $\frac{1}{2}\sigma$) a $2:1$ ratio of reservoir to sample may be expected to remove around 90% of an initial bias.

Any difference in the population standard deviations is also relevant. The percent-bias-removed values are higher than those shown in the table if $\sigma_1 < \sigma_2$ and lower if $\sigma_1 > \sigma_2$, again because of the problem of matching the highest members of sample 1.

Rubin also investigated the two variants of "nearest available" matching: (1) High–low, in which the members of sample 1 are ranked from high to low instead of at random—the highest member of sample 1 paired first, and so forth. (2) Low–high, in which the pairing proceeds from the lowest to the highest member of sample 1. For $\mu_1 > \mu_2$ Rubin found the "low–high" method best, the random method second best, and the "high–low" method third best, although the differences in performance were not great.

As with caliper matching, "nearest available" matching removed a somewhat higher percentage of the initial bias when $\sigma_2^2 > \sigma_1^2$ and removed a

Table 5.6.1.   Percent Reduction in Bias of $\bar{x}_1 - \bar{x}_2$ with "Nearest Available" Matching

n_1	$\dfrac{N_2}{n_1}$	$\dfrac{B}{\sigma} = \dfrac{1}{4}$	$\dfrac{1}{2}$	$\dfrac{3}{4}$	1
50	2	92	86	77	67
	3	96	94	90	84
	4	98	97	96	89
100	2	94	89	79	68
	3	98	96	92	85
	4	99	98	96	91

lower percentage when $\sigma_2^2 < \sigma_1^2$. With $\sigma_2^2 < \sigma_1^2$, the matches for the highest members of sample 1 tend to be too low, since population 1 has both a higher mean and a higher variance than population 2.

5.7 EFFECT OF MEAN MATCHING ON x

Table 5.7.1, which has the same format as Table 5.6.1, shows that the computer method of mean matching is highly successful in removing the bias of $\bar{x}_1 - \bar{x}_2$, as might be expected. The results in Tables 5.6.1 and 5.7.1 were obtained by computer simulation with x normal.

5.8 EFFECTS ON BIAS OF $\bar{y}_1 - \bar{y}_2$

The preceding results on the effects of different matching techniques on a confounding x variable were given as a step toward examining the effects of matching on the response variable y. These effects depend on the nature of the regression of y on x. First, consider bias with two populations. Several cases may be distinguished:

1.   *Linear Regression—The Same in Both Populations.*   With a single x and y continuous, the model is

$$y_{1j} = \alpha + \delta + \beta x_{1j} + e_{1j}; \qquad y_{2j} = \alpha + \beta x_{2j} + e_{2j}$$

where the constants α and β that define the regression are the same in both populations, and δ represents a constant effect of the difference in treatment. Since

$$E(\bar{y}_1 - \bar{y}_2) = \delta + \beta(\mu_1 - \mu_2)$$

Table 5.7.1.   Percent Reduction in Bias of $\bar{x}_1 - \bar{x}_2$ with Mean Matching

n_1	$\dfrac{N_2}{n_1}$	$\dfrac{B}{\sigma} = \dfrac{1}{4}$	$\dfrac{1}{2}$	$\dfrac{3}{4}$	1
50	2	100	99	91	77
	3	100	100	99	96
	4	100	100	100	100
100	2	100	100	96	80
	3	100	100	100	98
	4	100	100	100	100

it is clear that under this model the percent reduction in the bias of y equals that in x.

Next, suppose that y has a multiple linear regression on k variables, $x_1, x_2, \ldots, x_k$, to which matching has been applied. In this case

$$E(\bar{y}_1 - \bar{y}_2) = \delta + \sum_{i=1}^{k} \beta_i (\mu_{1i} - \mu_{2i})$$

The ratio of the final to the initial bias in y may be written

$$\frac{\sum_{i=1}^{k} \beta_i (\mu_{1i} - \mu_{2i})_m}{\sum_{i=1}^{k} \beta_i (\mu_{1i} - \mu_{2i})} \tag{5.8.1}$$

where $(\mu_{1i} - \mu_{2i})_m$ represents the difference in means for the ith x variable after matching. If the matching technique that is applied to each x produces the same percent reduction in bias, this is also the percent reduction in bias of y.

If the matching technique produces different percent reductions in bias for different x variables, the effect on y under multiple linear regression can be more complex. When all the terms $\beta_i (\mu_{1i} - \mu_{2i})$ have the same sign, the percentage of bias removed by matching lies between the least and greatest percentages for the individual x's, as follows from (5.8.1). However, if the terms $\beta_i (\mu_{1i} - \mu_{2i})$ have different signs, it is easy to construct cases in which the initial bias in y is small, because of cancellation of signs, and is increased by matching.

2.   *Nonlinear Regression—the Same in Both Populations.*   If the regression of y on x is $\phi(x)$, we are concerned with the effect of matching methods on the mean of $\phi(x_{1j}) - \phi(x_{2j})$. Some investigation has been made of monotone, moderately curved regressions such as $c_1 x + c_2 x^2$, for c_1, c_2, and x all positive and $e^{x/2}$ or $e^{-x/2}$.

In these cases the condition $\sigma_1 = \sigma_2$ (the variance of x is the same in the two populations) becomes important. With $\sigma_1 = \sigma_2 = \sigma$, consider $\phi(x) = c_1 x + c_2 x^2$, where x follows $N(\mu + \frac{1}{2}, 1)$ in population 1 and $N(\mu, 1)$ in population 2, with $\mu \geqslant 4$ so that negative values of x are rare. With either within-class or caliper matching, the percent reductions in $E[\phi(x_{1j})] - E[\phi(x_{2j})]$ are close to those in $E(x_{1j}) - E(x_{2j})$. These results are also suggested by the following algebraic argument. Let

$$y_{ij} = c_0 + c_1 x_{ij} + c_2 x_{ij}^2 + e_{ij} \qquad (i = 1, 2)$$

Then, apart from any treatment effect, the initial bias in $\bar{d} = \bar{y}_1 - \bar{y}_2$ is

$$E(\bar{d}) = c_1 (\mu_1 - \mu_2) + c_2 (\mu_1^2 + \sigma_1^2 - \mu_2^2 - \sigma_2^2) \tag{5.8.2}$$

where x_{ij} has mean μ_i and variance σ_i^2. With $\sigma_1^2 = \sigma_2^2$ this becomes

$$E(\bar{d}) = (\mu_1 - \mu_2)[c_1 + c_2(\mu_1 + \mu_2)]$$

The proportionate effect of matching on $E(\bar{d})$ will therefore equal its effect on $\mu_1 - \mu_2$, except that in (5.8.2) the value of $\mu_1 + \mu_2$ is slightly altered by matching and the variances σ_{1m}^2 and σ_{2m}^2 will not be exactly equal in the population created by matching.

However, when $\sigma_1 \neq \sigma_2$, the percentage reduction in bias due to matching depends on the size and sign of the term $c_2(\sigma_1^2 - \sigma_2^2)$ in the initial bias, and on how matching affects population variances. Taking $y = x^2$, with $\mu_1 = 4.5$ and $\mu_2 = 4.0$, within-class matching with two classes reduces the bias by 63% when $\sigma_1^2 = \sigma_2^2 = 1$; by 70% when $\sigma_1^2 = \frac{2}{3}$ and $\sigma_2^2 = \frac{4}{3}$; and by only 52% when $\sigma_1^2 = \frac{4}{3}$ and $\sigma_2^2 = \frac{2}{3}$. Some results for within-class matching, with $y = e^{x/2}$, $y = e^{-x/2}$, and $\mu_1 - \mu_2 = 0.5$, are given in Table 5.8.1.

When $\sigma_1 = \sigma_2$, results for the effect of within-class matching on x can apparently be used as a rough, if slightly optimistic, guide to its effect on monotone, moderately curved regressions. This is not so when there are substantial differences in variances. Rubin's results for "nearest available" matching will be given in Chapter 6 for comparison with regression adjustments.

Mean matching is highly successful in diminishing $\mu_1 - \mu_2$ in the matched samples and is essentially intended to cope with a linear regression of y on x. Its performance under nonlinear regressions will depend both on the nature of the regression and on the specific method of mean matching. Rubin's method, for instance, will tend to make σ_2^2 less than σ_1^2 in the matched populations even if they are initially equal, since it chooses values of x_{2j} near to $\bar{x}_1$. Rubin's method is likely to do poorly on regressions like $e^{x/2}$, even if $\sigma_1 = \sigma_2$ initially, and should be avoided if nonlinear regressions are suspected; other matching methods are preferable to nearest available pair matching.

Table 5.8.1. Percent Reductions in Bias of y for Within-Class Matching When $y = x$, $y = e^{x/2}$ and $y = e^{-x/2}$

$E(y\|x)$ $(\sigma_{1x}^2, \sigma_{2x}^2)$	x $(1,1)$	$e^{x/2}$ $(1,1)$	$(\frac{2}{3},\frac{4}{3})$	$(\frac{4}{3},\frac{2}{3})$	$e^{-x/2}$ $(1,1)$	$(\frac{2}{3},\frac{4}{3})$	$(\frac{4}{3},\frac{2}{3})$
Two Classes	64	61	94	44	61	47	88
Three Classes	79	76	105[a]	60	76	64	110[a]
Four Classes	86	84	108[a]	69	84	73	107[a]

[a]Entry 105 denotes that remaining bias is 5% of the original bias, but of opposite sign. Other entries exceeding 100 are similarly interpreted.

3. *Regression—Different in the Two Populations.* The concept of matching is geared to the assumption that the regression of y on x is the same in the populations being compared. Suppose that there are different linear regressions in the two populations, namely,

$$y_{1j} = \alpha_1 + \delta + \beta_1 x_{1j} + e_{1j}; \qquad y_{2j} = \alpha_2 + \beta_2 x_{2j} + e_{2j} \qquad (5.8.3)$$

where, as usual, δ is the effect of the difference in treatment. Then

$$E(\bar{y}_1 - \bar{y}_2) = \delta + (\alpha_1 - \alpha_2) + \beta_1 \mu_1 - \beta_2 \mu_2 \qquad (5.8.4)$$

Matching will not affect the term $\alpha_1 - \alpha_2$, which represents a constant bias. Further, even if we succeed in making $\mu_1 = \mu_2 = \mu$, a bias $(\beta_1 - \beta_2)\mu$ remains after matching. It is best to avoid matching in this case.

Unfortunately, it is possible only in a before–after study to detect from the data that this situation exists before deciding whether to match. In such a study, y and x are measured in two populations *before* any difference in treatment has occurred, and also *after* a period of exposure to the two treatments. From the "before" data, the regressions of y on x in the two populations can be estimated and compared at a time when the populations have no difference in treatment. The existence of a model like (5.8.3) can thus be detected before a decision on matching is made. In an "after only" study, the measurement of y is usually postponed until after the samples have been selected (i.e., the decision to match or not to match has already been made).

In the final results in an "after only" study, the finding of different linear regressions in the two populations has another possible interpretation. Assuming $\alpha_1 = \alpha_2$, from (5.8.3) it follows that for a given value of x

$$y_1 - y_2 = \delta + (\beta_1 - \beta_2)x + e_1 - e_2 \qquad (5.8.5)$$

This relation might hold because the effect of the difference in treatments is $\delta + (\beta_1 - \beta_2)x$, varying with the level of x. If we have matched on x, a linear regression of $\bar{y}_1 - \bar{y}_2$ on x would reveal this situation. In fact, in many studies the investigator expects the effect of the difference in treatments to vary with x. Of course the investigator could be misled in this interpretation if the regressions actually have different slopes in the two populations.

5.9   EFFECT OF MATCHING ON THE VARIANCE OF $\bar{y}_1 - \bar{y}_2$

Let us now consider how matching increases precision when there is no danger of bias. Consider a linear regression of y on x, which is the same in both populations, with $\mu_1 = \mu_2$. For this,

$$V(\bar{y}_1 - \bar{y}_2) = \beta^2 V(\bar{x}_1 - \bar{x}_2) + V(\bar{e}_1 - \bar{e}_2) \qquad (5.9.1)$$

Table 5.9.1. Percent Reduction ($f\rho^2$) in $V(\bar{y}_1 - \bar{y}_2)$ due to Matching in the "No Bias" Situation (Linear Regression)

Number of Classes	f	0.3	0.4	0.5	ρ 0.6	0.7	0.8	0.9
2	0.64	6	10	16	23	31	41	52
3	0.81	7	13	20	29	40	52	66
4	0.88	8	14	22	32	43	56	71
5	0.92	8	15	23	33	45	59	75
∞	1.00	9	16	25	36	49	64	81

With random (unmatched) samples of size n, the two terms on the right-hand side can be written in terms of the correlation ρ between y and x.

$$V(\bar{y}_1 - \bar{y}_2) = \frac{2}{n}\left[\rho^2\sigma_y^2 + \left(1 - \rho^2\right)\sigma_y^2\right]$$

If the fractional reduction in $V(\bar{x}_1 - \bar{x}_2)$ due to matching is f, this reduction affects the component $\rho^2\sigma_y^2$, but not the residual component $(1 - \rho^2)\sigma_y^2$. Hence the fractional reduction in $V(\bar{y}_1 - \bar{y}_2)$ due to matching is $f\rho^2$, and can be calculated for a given ρ from the values of f in preceding tables. Table 5.9.1 shows the percent reductions from within-class matching for the smaller numbers of classes.

With three or more classes, the percent reductions are determined primarily by the value of ρ rather than by the number of classes. Matching for increased precision when there is no danger of bias, does not begin to pay substantial dividends until ρ is 0.5 or greater.

When a decision about matching is to be made, either for protection against bias or for increased precision, an important question to consider is "What are the alternatives to matching? The principal alternatives—adjustments during the statistical analysis—are the subject of Chapter 6, which includes comparisons with matching where available.

5.10 INTRODUCTION TO STATISTICAL ANALYSIS OF PAIR-MATCHED SAMPLES

In this section we introduce methods of statistical analysis for pair-matched (caliper or "nearest available") and mean-matched samples. For within-class matching the methods of analysis are essentially the same as those for the adjustment of unmatched samples, and will be discussed in Chapter 6.

Pair Matching: y Continuous

The data form a two-way classification (treatments and pairs or matched groups). Under the additive model

$$y_{ij} = \mu + \tau_i + \gamma_j + e_{ij}$$

with $V(e_{ij}) = \sigma^2$, the usual two-way analysis of variance provides an estimate of the standard error of any comparison among the treatment means. With only two treatments, we may equivalently analyze the column of differences $d_j = y_{1j} - y_{2j}$, between the members of a pair, in order to estimate the standard error of $\bar{d} = \bar{y}_1 - \bar{y}_2$. Note that the analysis with two treatments does not require the assumption that $\sigma_1^2 = \sigma_2^2$. Similarly, if the variances $V(e_{ij}) = \sigma^2$ are thought to change from treatment to treatment, valid standard errors and t tests for any comparison $\Sigma \lambda_i \bar{y}_i$ are obtained by analyzing the column of values $\Sigma \lambda_i y_{ij}$.

The analysis of matched pairs is usually directed at estimation and testing of the mean difference $\bar{d} = \bar{y}_1 - \bar{y}_2$. However, with tight matching it is also possible to examine whether $d_j = y_{1j} - y_{2j}$ varies with the level of x. One approach is to let $x_j = (x_{1j} + x_{2j})/2$ and compute the linear regression of d_j on x_j, which constitutes 1 d.f. (degree of freedom) from the $(n - 1)$ d.f. for the variation of d_j from pair to pair. Higher polynomial regression terms may be added if appropriate, or multivariate regression of d_j may be used on different x variables than were used in matching.

Pair Matching: y (0, 1)

If y represents a two-way classification with two pair-matched treatments, a member of any pair can only have the y values 0 or 1. Thus the pairs have only the four y values (1, 1), (1, 0), (0, 1), and (0, 0), where the first number refers to treatment 1 (T_1) and the second number to treatment 2 (T_2). The data may be summarized as follow:

T_1	T_2	Number of Pairs
1	1	n_{11}
1	0	n_{10}
0	1	n_{01}
0	0	n_{00}
Total		n

The proportions of "ones" for the two treatments are $\hat{p}_1 = (n_{11} + n_{10})/n$ and $\hat{p}_2 = (n_{11} + n_{01})/n$. Thus $\bar{d} = (n_{10} - n_{01})/n$. As McNemar (1947) and other investigators have shown, the null hypothesis $p_1 = p_2$ is tested by regarding n_{10} and n_{01} as binomial successes and failures from $n_{10} + n_{01}$ trials, with probability of success $\frac{1}{2}$ on the null hypothesis. An exact test can be made from the binomial tables. For an approximate test, the value of χ^2 corrected for continuity with 1 d.f. is

$$\frac{(|n_{10} - n_{01}| - 1)^2}{n_{10} + n_{01}}$$

Stuart (1957) gives the estimated standard error of $\bar{d} = \hat{p}_1 - \hat{p}_2$ as

$$\left(\frac{\hat{p}_{10} + \hat{p}_{01} - (\hat{p}_{10} - \hat{p}_{01})^2}{n} \right)^{1/2}$$

Billewicz (1965) reports that in 9 out of 20 examples of matching in the field of medicine, the analysis used was (incorrectly) that appropriate to independent rather than matched samples. This mistake overestimates the standard error of $\bar{d}$ and underestimates χ^2. [See Cochran (1950) for an extension of the pair-sample methods when more than two treatments are used.]

With two treatments, let p_{1j} and p_{2j} be the true probabilities of success in the jth pair. In pairing, we presumably expect the probabilities of success to vary from pair to pair. In seeking a model that describes how p_{1j} and p_{2j} vary from pair to pair, many authors [writing q for $(1 - p)$] have used the following relations:

$$\frac{p_{1j}}{q_{1j}} = \psi \lambda_j; \qquad \frac{p_{2j}}{q_{2j}} = \lambda_j \qquad\qquad (5.10.1)$$

where λ_j represents the level of the jth pair and ψ measures the disparity between the effects of the treatments. In model (5.10.1), the quantity that is regarded as constant from pair to pair is $\psi = p_{1j}q_{2j}/p_{2j}q_{1j}$, sometimes called the *odds ratio*, rather than $\delta = p_{1j} - p_{2j}$. The model assumes that the effects of the treatment and the pair are additive on the scale of $\log(p_{ij}/q_{ij})$, called the "logit of p_{ij}." An additive model on the scale of p_{ij} itself has the logical difficulty that p_{ij} must lie between 0 and 1.

With model (5.10.1) the quantity to be estimated is the odds ratio ψ. If the λ_j are regarded as nuisance parameters, D. R. Cox (1958) has shown that (1) an optimum estimate of ψ uses the (1,0) and (0, 1) pairs only, and that (2) the ratio $n_{10}/(n_{10} + n_{01})$ is a binomial estimate of $\theta = \psi/(1 + \psi)$, based on a sample of size $n_{10} + n_{01}$. This result provides an estimate of θ

and hence of $\psi = \theta/(1 - \theta)$. Similarly, confidence limits for ψ can be obtained from binomial confidence limits for θ as shown in Hald's (1952) tables or the Fisher and Yates tables (1953).

5.11 ANALYSIS WITH MEAN MATCHING; y CONTINUOUS

So far as I know, analysis with mean matching is seldom used. It has merit when the regression of y on x is linear with the same slope in both populations. The following analysis makes these assumptions. Let

$$y_{1u} = \mu + \tau_1 + \beta x_{1u} + e_{1u}; \qquad y_{2v} = \mu + \tau_2 + \beta x_{2v} + e_{2v}$$

Hence

$$\bar{y}_1 - \bar{y}_2 = \tau_1 - \tau_2 + \beta(\bar{x}_1 - \bar{x}_2) + \bar{e}_1 - \bar{e}_2$$

An effective mean matching makes $\bar{x}_1 - \bar{x}_2$ so close to zero that $\bar{y}_1 - \bar{y}_2$ serves as the estimate of $\tau_1 - \tau_2$, with variance

$$V(\bar{y}_1 - \bar{y}_2) = \frac{\sigma_{1e}^2 + \sigma_{2e}^2}{n}$$

In order to estimate σ_{1e}^2 and σ_{2e}^2 and hence $V(\bar{y}_1 - \bar{y}_2)$, the effects of the linear regression of y on x must be removed from variations in y. Let $(yy)_1$ denote $\Sigma(y_{1u} - \bar{y}_1)^2$, etc. Then

$$\hat{\sigma}_{1e}^2 = \frac{(yy)_1 - (yx)_1^2/(xx)_1}{n - 2}$$

with a similar expression for $\hat{\sigma}_{2e}^2$.

A large-sample $1 - \alpha$ confidence interval for $\tau_1 - \tau_2$, with effective mean matching, where y has a linear regression on x with the same slope in both populations is thus

$$\bar{y}_1 - \bar{y}_2 - z_{\alpha/2}\sqrt{\frac{\hat{\sigma}_{1e}^2 + \hat{\sigma}_{2e}^2}{n}} < \tau_1 - \tau_2 < \bar{y}_1 - \bar{y}_2 + z_{\alpha/2}\sqrt{\frac{\hat{\sigma}_{1e}^2 + \hat{\sigma}_{2e}^2}{n}}$$

5.12 SUMMARY

In an observational study systematic differences between the populations from which different treatment groups are drawn can have two effects on comparisons $\bar{y}_1 - \bar{y}_2$ between the response means for samples exposed to different treatments. These differences may create a bias in these comparisons and may decrease precision of comparisons. In attempting to avoid these consequences the first step is to list the principal variables x that influence y and are not themselves affected by the treatments. Such variables are called *confounding variables* (sometimes called "covariates" or "control variables"). They may be classifications, or discrete or continuous variables.

Matching is a common method of handling such confounding x variables at the planning stage by making the samples for different treatments resemble each other in certain respects. The matching method used depends on the nature of the x distributions.

If the x's are classifications, the cells that are created by this multiple classification are formed. In *within-class* or *frequency matching*, each member of any sample has a partner in any other sample belonging to the same cell, so that in the ith cell all samples for different treatments contain the same number of members n_i. This method is often used with discrete or classified x variables (e.g., number of children, age) by first grouping the values of the variable into classes.

With continuous or discrete x variables *caliper matching* requires the x values for partners in different samples to agree within prescribed limits $\pm a$. *Mean Matching* concentrates on making the means $\bar{x}_{ti}$ for different treatments agree as closely as possible for the ith x variable.

The idea of matching is simple to understand. Its objective is to free the comparisons of the means $\bar{y}_t$ from the effects of differences among the x distributions in different treatment groups. Perfect matching removes from these comparisons the effects of any shape of relationship between y and the x's, provided that this shape is the same in all populations being compared. Hence the statistical analysis of matched samples is relative simple.

The primary disadvantage is the time and effort required to construct matched samples. The degree of difficulty depends on the desired sample size, the sizes of the reservoirs available for seeking matches, the number of x's to be matched, the tightness of the matching rules (caliper matching, in general is more difficult than within-class matching) and the sizes of the differences between the x distributions in different populations. A case-by-case search for paired matches on say four x variables can be tedious and may require several relaxations of the original matching rules in order to find matches. One consequence of matching whose effects are more difficult to assess is that the sample-population relationship is disturbed. In matched sampling every sample may be a nonrandom sample from its own population.

In finding matches, computers should be able to take over much of the work if the x's in the reservoirs are in a form suitable for data entry. For instance, using within-class matching on three x variables having c_1, c_2, and c_3 classes, the computer can arrange and print the data in the $c_1 \times c_2 \times c_3$ cells from which matches are easily selected. A similar method is a considerable help in seeking caliper matches.

Sometimes an investigator needs all n cases available from one population and has a limited reservoir from the second population. With x continuous it is not clear whether caliper matches with a prescribed $\pm a$ can be found for all n. For this problem D. Rubin has developed a method of computer matching, called *nearest available* matching, that guarantees that every case is matched.

In comparing two samples, a primary objective of matching is, of course, to remove bias in $\bar{y}_1 - \bar{y}_2$ due to differences in the x distributions. If y has the same linear regression on x in both populations, the percentage reduction in the bias of $\bar{y}_1 - \bar{y}_2$ due to matching equals the percentage reduction in the bias of $\bar{x}_1 - \bar{x}_2$. Consequently, the effect of matching on the bias of $\bar{x}_1 - \bar{x}_2$ is examined first.

In within-class matching, some types of classified x's are such that two members of the same class are identical with regard to x. In this event, within-class matching completely removes any initial bias in $\bar{x}_1 - \bar{x}_2$. But many classified x's (e.g., social level, degree of aggressiveness) more nearly represent a grouping of an underlying continuous x, as is the case when a continuous x is deliberately grouped in order to use within-class matching. In this situation, matching with two, three, four and five classes removes approximately 64%, 80%, 87%, and 91% of an initial bias in $\bar{x}_1 - \bar{x}_2$.

With x continuous, the effect of caliper matching, to within $\pm a$ units, depends primarily on the ratio a/σ_x and to some extent on the way in which the caliper matches are constructed. "Loose" caliper matching to within $\pm 0.9\sigma_x$ removes slightly more than 80% of an initial bias (as effective as within-class matching with three classes). Caliper matching to within $\pm 0.4\sigma_x$ removes 95–97% of the initial bias.

Mean matching is highly successful in removing bias in $\bar{x}_1 - \bar{x}_2$ even if all n members of sample 1 must be used and the sample 2 reservoir is only of size $2n$.

The effect of "nearest available" matching depends on the size of the initial bias $\bar{x}_1 - \bar{x}_2$ and on the size of the reservoir in population 2. With a reservoir of size $4n$ this method removes nearly all the initial bias, unless this bias is exceptionally large. Even a reservoir of size $2n$ should remove around 90% of a moderate-sized initial bias.

Suppose that the regression of y on x is the same in two populations but is nonlinear, being monotone and moderately curved as represented by a quadratic regression or by $y = e^{\pm x/2}$. Some evidence indicates that for within-class and caliper matching, the percent reduction in the bias of

$\bar{y}_1 - \bar{y}_2$ is only slightly less than that in the linear case, provided x has the same variance in the two populations. When σ_{1x}^2 and σ_{2x}^2 are unequal this result does not hold; the percentage reduction in bias is sometimes much greater and sometimes much less than when $\sigma_{1x}^2 = \sigma_{2x}^2$. Mean matching should be avoided when the regression of y on x is curved.

Matching methods are only partially successful to varying degrees in removing an initial bias in $\bar{y}_1 - \bar{y}_2$ due to confounding x variables. Under a linear regression of y on x, removal of over 90% of an initial bias requires five classes in within-class matching, or caliper matching to within $\pm 0.4\sigma_x$. Matching is not suitable when the regression of y on x is of a different form in the two populations.

Matching is also used to increase the precision of the comparison $\bar{y}_1 - \bar{y}_2$ when the x distribution is thought to be the same in the two populations, that is, where there is no danger of bias. For within-class and caliper matching, the percent reduction f in $V(\bar{x}_1 - \bar{x}_2)$ is similar to that in the bias of $\bar{x}_1 - \bar{x}_2$. Under a linear regression the percent reduction in $V(\bar{y}_1 - \bar{y}_2)$ is $f\rho^2$, where ρ is the correlation between y and x. Thus the reduction in $V(\bar{y}_1 - \bar{y}_2)$ does not become substantial until $|\rho|$ exceeds 0.5.

REFERENCES

Billewicz, W. Z. (1965). The efficiency of matched samples: an empirical investigation. *Biometrics*, **21**, 623–644.

Cochran, W. G. (1950). The comparison of percentages in matched samples. *Biometrika*, **37**, 256–266 [Collected Works #43].

Cochran, W. G. (1968). The effectiveness of adjustment by subclassification in removing bias in observational studies. *Biometrics*, **24**, 295–313 [Collected Works #90].

Cox, D. R. (1957). Note on grouping. *J. Am. Statist. Assoc.*, **52**, 543–547.

Cox, D. R. (1958). Two further applications of a model for binary regression. *Biometrika*, **45**, 562–565.

Douglas, J. W. B. (1960). 'Premature' children at primary schools. *Br. Med. J.*, **1**, 1008–1013.

Feller, W. (1957). *An Introduction to Probability Theory and its Applications* (2nd ed.). Wiley, New York.

Fisher, R. A. and F. Yates (1953). *Statistical Tables for Biological, Agriculture and Medical Research* (4th ed.). Oliver and Boyd, Edinburgh, Scotland.

Hald. A. (1952). *Statistical Tables and Formulas*. Table XI. Wiley, New York.

McNemar, Q. (1947). Note on the sampling error of the difference between correlated proportions or percentages. *Psychometrika*, **12**, 153–157.

Ogawa, J. (1951). Contributions to the theory of systematic statistics. I. *Osaka Math. J.*, **3**, 175–213.

Rubin, D. B. (1970). The Use of Matched Sampling and Regression Adjustment in Observational Studies. Ph.D. Thesis, Harvard University, Cambridge, Mass.

Rubin, D. B. (1973a). Matching to remove bias in observational studies. *Biometrics*, **29**, 159–183.

Rubin, D. B. (1973b). The use of matched sampling and regression adjustment to remove bias in observational studies. *Biometrics*, **29**, 185–203.

Stanley, J. C. (1966). A common class of pseudo-experiments. *Am. Educational Res. J.*, **3**, 79–87.

Stuart, A. (1957). The comparison of frequencies in matched samples. *Br. J. Statist. Psychol.*, **10**, 29–32.

CHAPTER 6

Adjustments in Analysis

6.1 INTRODUCTION

When the principal confounding x variables have been noted, the main alternative to matching, as discussed in Section 5.1, is to make adjustments in the course of the statistical analysis. The objectives remain the same—to protect against bias and to increase the precision of the comparision between treatment means or proportions.

In some situations an adjustment method is the only possibility because matching is not feasible or is obviously unattractive. The economics of the study may require that the x's and y be measured simultaneously, after the samples have been chosen, so that advance creation of matched samples is ruled out. As mentioned, matching is confined mainly to smaller-sample and simpler studies, often two-group comparisons. Matching becomes troublesome with large samples, when subjects enter the study only over an extended time period, and also as the number of treatments to be compared or the number of variables to be matched increases.

This chapter describes the principal methods of adjustment in the simpler situations. Where possible, comparisons of the performance of matching and adjustment methods will be noted, since many studies could use either method. Once again, the details of the adjustment method depend on the scales in which y and the x's are measured.

6.2 y CONTINUOUS: x's CLASSIFIED

With two populations we assume that the adjustment method starts with independent random samples. Having selected the classified x variables for which adjustment is to be made, we first arrange the data from the two

102

samples into the cells created by this classification. Let the sample numbers in the ith cell of the classification be n_{1i} and n_{2i}. The response means are $\bar{y}_{1i}$ and $\bar{y}_{2i}$ and the proportions (if y is 0 and 1) are p_{1i} and p_{2i}. The only difference between this situation and within-class matching is that in the latter, $n_{1i} = n_{2i} = n_i$ in every cell.

If $d_i = \bar{y}_{1i} - \bar{y}_{2i}$, the estimates of the overall treatment difference for the two methods are

$$\bar{d} = \bar{y}_1 - \bar{y}_2 = \sum \frac{n_i}{n} d_i \qquad \text{(matched samples)}$$

and

$$\bar{d}_a = \bar{y}_{1a} - \bar{y}_{2a} = \sum W_i d_i \qquad \text{(random samples)}$$

The matched-sample weights, n_i/n, weight each class mean by its class size, resulting in simply the difference between two means; the weights W_i, with $\sum W_i = 1$, are chosen by the investigator; and the subscript a denotes adjustment. In both methods any remaining bias arises from the fact that $E(d_i) \neq 0$ when the underlying confounding x variables have different distributions in the two populations. If $E(d_i)$ were constant from cell to cell, matching and adjustment would be equally effective in reducing bias for any choice of weights W_i. In fact, in the presence of bias, $E(d_i)$ varies from cell to cell and is usually greater at the high and low extremes of the x distributions than near the medians. However, for the weights likely to be used in practice, within-class matching and adjustment may be regarded as roughly equally effective in reducing bias. Matching, though, has the advantage of a simpler estimate $\bar{y}_1 - \bar{y}_2$ that avoids weighting.

We now consider the choice of weights. Suppose first that the mean difference $\delta = \tau_1 - \tau_2$ between the effects of the two treatments is the same in every cell. It follows that in random samples *any* weighted mean $\sum W_i d_i$ is an estimate of δ, apart from within-cell bias in d_i. If this situation holds, the choice of weights may be determined by considering convenience or statistical precision. If, however, $\tau_{1i} - \tau_{2i} = \delta_i$ varies from cell to cell, $\sum W_i d_i$ becomes an estimate of $\sum W_i \delta_i$, a quantity whose value now depends on the choice of weights. It may be clear on inspection of the data, particularly with large samples, that there are real differences between the effects of the treatments and that these differences vary from cell to cell. Section 6.4 discusses this further. For now, we assume δ constant and discuss the estimation and testing of δ.

If σ_{1i}^2 and σ_{2i}^2 are the two population variances within the ith cell,

$$V(\bar{d}_a) = V\left(\sum W_i d_i\right) = \sum W_i^2 \left(\frac{\sigma_{1i}^2}{n_{1i}} + \frac{\sigma_{2i}^2}{n_{2i}}\right) \qquad (6.2.1)$$

This variance is minimized by taking W_i proportional to

$$w_i = \frac{1}{\sigma_{1i}^2/n_{1i} + \sigma_{2i}^2/n_{2i}}$$

Of course, $W_i = w_i/\Sigma w_i$. The resultant minimum variance is $1/\Sigma w_i$.

These general formulas are needed when assumptions of equality of within-class variances cannot be made. If all n_{1i} and n_{2i} exceed 30, choice of w_i inversely proportional to the estimated variance

$$\hat{w}_i = \frac{1}{s_{1i}^2/n_{1i} + s_{2i}^2/n_{2i}} \tag{6.2.2}$$

should do almost as well, where $V(\bar{y}_{1a} - \bar{y}_{2a}) \doteq 1/\Sigma \hat{w}_i$ [Meier (1953)].

Various particular cases arise when assumptions can be made about the within-cell variances. In the formulas below, s_i^2, s_1^2, s_2^2, and s^2 are the appropriate pooled estimates of σ_i^2, σ_1^2, σ_2^2, and σ^2 respectively

$$\sigma_{1i}^2 = \sigma_{2i}^2 = \sigma_i^2; \quad \hat{w}_i = \frac{n_{1i}n_{2i}}{(n_{1i} + n_{2i})s_i^2} \tag{6.2.3}$$

$$\sigma_{1i}^2 = \sigma_1^2; \sigma_{2i}^2 = \sigma_2^2; \quad \hat{w}_i = \frac{n_{1i}n_{2i}}{n_{2i}s_1^2 + n_{1i}s_2^2} \tag{6.2.4}$$

and

$$\sigma_{1i}^2 = \sigma_{2i}^2 = \sigma^2; \quad \hat{w}_i = \frac{n_{1i}n_{2i}}{(n_{1i} + n_{2i})s^2} \tag{6.2.5}$$

In this case [(6.2.5)] it is customary to take $w_i' = \hat{w}_i s^2 = n_{1i}n_{2i}/(n_{1i} + n_{2i})$ as relative weights. The variance of $\Sigma w_i' \bar{d}_i/\Sigma w_i'$ is

$$\frac{s^2}{\Sigma w_i'} = \frac{s^2}{\Sigma[n_{1i}n_{2i}/(n_{1i} + n_{2i})]}$$

Formula (6.2.5) is often used when the σ_{ti}^2 are thought not to vary much.

When the σ_{ti}^2 vary little and the two total sample sizes are equal, the simpler weights, w_i proportional to $n_{1i} + n_{2i} = n_i$, the combined sample size in cell i, seldom do much worse than the optimum weights $n_{1i}n_{2i}/n_i$ in (6.2.5) for this case. With weights proportional to n_i and with σ_{ti}^2 constant, for instance,

$$\hat{V}(\bar{y}_{1a} - \bar{y}_{2a}) = \frac{s^2\Sigma(n_i^3/n_{1i}n_{2i})}{(\Sigma n_i)^2} \qquad (6.2.6)$$

In general, no simple statement can be made about the relative precision of the comparison $\bar{y}_1 - \bar{y}_2$ for matched samples and $\bar{y}_{1a} - \bar{y}_{2a}$ for adjusted random samples both of total size n, because this depends on the way in which the σ_{ti}^2's vary from cell to cell and on the choice of weights. However, this comparison is of interest when any underlying x has the same distribution in the two populations. In this situation the purpose of matching or adjustment is to increase precision, because there is no danger of bias. Any difference in variances of $\bar{y}_1 - \bar{y}_2$ and $\bar{y}_{1a} - \bar{y}_{2a}$ is likely to be minor because differences between n_{1i} and n_{2i} in any cell will arise only from random-sampling variation. If $n_{1i} = n_{2i} = n_i/2$, then the variance of (6.2.6) becomes $4s^2/\Sigma n_i$ as does the variance derived from the conditions of (6.2.5).

6.3 y BINOMIAL: x's CLASSIFIED

The adjusted mean difference with two random samples is of the form $\Sigma w_i(\hat{p}_{1i} - \hat{p}_{2i})$, where the $\hat{p}_{ti}$ are the observed proportions of "ones" (successes) in cell i. As with y continuous, the difference in effectiveness of matching and adjustment depends on the within-cell biases and the choice of the w_i. The difference should be small for most choices of the w_i in practice.

In choosing the w_i for an initial test of significance or estimation of an overall difference (ignoring within-cell bias), some theoretical issues have to be considered. As indicated in Section 5.10, the assumption of an additive model of the form

$$p_{1i} = \mu + \tau_1 + \gamma_i; \qquad p_{2i} = \mu + \tau_2 + \gamma_i$$

is unreasonable on logical grounds, since the p_{ti} must lie between 0 and 1. Most recent work on the analysis of proportions in multiple classifications has assumed an additive model in the scale of $\log(p_{ti}/q_{ti})$, where $q_{ti} = 1 - p_{ti}$. In this situation the investigator may still wish to estimate an overall difference $p_1 - p_2$, particularly if it is not clear that there is a real difference in treatment effects, so that a test of significance is desired.

Under an additive model in the logit scale, Cochran (1954) has shown that an effective choice is to take w_i proportional to $n_{1i}n_{2i}/n_i$, where $n_i = n_{1i} + n_{2i}$. An approximate test of significance of the null hypothesis,

$p_{1i} = p_{2i}$ (for all i), is made as follows. [This test does not fully address the null hypothesis stated, but uses its assumptions. The test is directed to the more-general null hypothesis $\Sigma w_i(p_{1i} - p_{2i}) = 0$, which includes the stated null hypothesis.] With $w_i = n_{1i}n_{2i}/n_i$, the weighted difference $\Sigma w_i(\hat{p}_{1i} - \hat{p}_{2i})$ has approximate estimated variance

$$\sum w_i \hat{p}_i \hat{q}_i$$

where $\hat{p}_i$ is the overall proportion of successes in cell i. The test is made by treating

$$\frac{\Sigma w_i(\hat{p}_{1i} - \hat{p}_{2i})}{(\Sigma w_i \hat{p}_i \hat{q}_i)^{1/2}}$$

as a normal deviate.

Two refinements by Mantel and Haenszel (1959), who also developed this test from a different viewpoint, are worth using when some n_{ti} are small, as often occurs. One technique involves inserting a correction for continuity; the other uses a slightly different variance formula. In this form the normal deviate for a test of significance is taken as

$$\frac{|\Sigma w_i(\hat{p}_{1i} - \hat{p}_{2i})| - \frac{1}{2}}{[\Sigma n_{1i}n_{2i}\hat{p}_i \hat{q}_i/(n_{1i} + n_{2i} - 1)]^{1/2}}$$

If we conclude that there is a real overall difference, the null hypothesis being rejected, and if we wish to attach a standard error to the weighted mean difference $\Sigma w_i(\hat{p}_{1i} - \hat{p}_{2i})/\Sigma w_i$, then we can no longer regard $p_{1i} = p_{2i}$. The estimated standard error of the weighted mean difference is

$$\left[\sum w_i^2 \left(\frac{\hat{p}_{1i}\hat{q}_{1i}}{n_{1i} - 1} + \frac{\hat{p}_{2i}\hat{q}_{2i}}{n_{2i} - 1}\right)\right]^{1/2} \bigg/ \sum w_i$$

6.4   TREATMENT DIFFERENCE VARYING FROM CELL TO CELL

In this situation the choice of weights should not be dictated by considerations of precision, particularly in large-sample studies in which any reasonable weighting gives adequate precision. Several situations in choosing weights may arise. The investigator may find that on trying several likely sets of weights, the estimates $\Sigma w_i(\bar{y}_{1i} - \bar{y}_{2i})$ or $\Sigma w_i(\hat{p}_{1i} - \hat{p}_{2i})$, while differing from set to set, agree sufficiently well that any conclusions to be drawn, or any action to be taken, would be the same. If the estimates disagree more widely, the choice of weights is more critical. This problem is old and familiar in vital statistics, for example, in the international comparision of overall death rates. One device used there is to take the weights from some standard population that is regarded as a target population. The process of adjustment is called *standardization*. To illustrate, Keyfitz (1966) reports that the mortality rate for French females in 1962 exceeded the rate for American females in 1963 by 14, 16, or 51%, according to the different standard populations used for weighting.

In the face of substantial differences between estimates based on different sets of weights, any overall estimate $\hat{\delta}$ may be, to some extent, arbitrary and liable to misinterpretation unless there is a specific target population with known weights for which an estimate of δ is clearly relevant. Otherwise, the most useful report on the data may be to summarize and try to interpret how d_i varies from cell to cell.

In both matched and independent samples, rough tests of the null hypothesis that the δ_i are the same in each cell are possible, and sometimes helpful. The simplest case is one in which y is continuous and the within-cell variance can be assumed constant. This is unlikely to be strictly true in observational studies, but might not be seriously wrong. Let s^2 denote the pooled within-cell variance and

$$w_i' = \frac{n_{1i}n_{2i}}{n_i} ; \qquad d_i = \bar{y}_{1i} - \bar{y}_{2i}; \qquad \bar{d}_a = \frac{\Sigma w_i' d_i}{\Sigma w_i'}$$

Calculate the weighted sum of squares

$$Q = \Sigma w_i' \left(d_i - \bar{d}_a \right)^2 = \Sigma w_i' d_i^2 - \frac{\left(\Sigma w_i' d_i\right)^2}{\Sigma w_i'} \tag{6.4.1}$$

With c classes, assuming normality, the quantity $Q/(c-1)s^2$ is distributed on the null hypothesis as F with $(c-1)$ and $(n_1 + n_2 - 2c)$ d.f. (degrees of freedom). Large values of F cause rejection of the null hypothesis.

With y binomial, and with y continuous, when within-cell variances vary, a method is available which leads to a large-sample χ^2 test. Let $d_i = \bar{y}_{1i} - \bar{y}_{2i}$, or $\hat{p}_{1i} - \hat{p}_{2i}$. Compute an unbiased estimate of the variance of d_i according to the assumptions that seem reasonable. In the general case,

$$\hat{V}(d_i) = \frac{s_{1i}^2}{n_{1i}} + \frac{s_{2i}^2}{n_{2i}} \qquad (y \text{ continuous})$$

and

$$\hat{V}(d_i) = \frac{\hat{p}_{1i}\hat{q}_{1i}}{n_{1i} - 1} + \frac{\hat{p}_{2i}\hat{q}_{2i}}{n_{2i} - 1} \qquad (y \text{ binomial})$$

Assume $\hat{w}_i = 1/\hat{V}(d_i)$. Then on the null hypothesis, $\Sigma \hat{w}_i(d_i - d_w)^2$ is approximately χ^2 with $(c - 1)$ d.f. where $d_w = \Sigma \hat{w}_i d_i$. [Sometimes assumptions such as those given in Eqs. (6.2.3), (6.2.4), and (6.2.5) seem natural and lead to slightly different formulas.]

The large-sample χ^2 test may be directed at more-specific alternatives by breaking χ^2, or the numerator Q of F in (6.4.1), into components. For instance, the c cells might subdivide into three sets, with reason to expect d_i to be constant within each set, but to vary from set to set. Then χ^2 is broken down into four components—one for "between sets" and one for "within each set." Similarly, if the cells represent an ordered classification, with scores z_i assigned to the cells, the test of the linear regression of d_i on z_i may be of interest. (Remember to take the different weights into account.) Thus in the linear-regression test for χ^2 with 1 d.f., we calculate

$$N = \sum \hat{w}_i d_i z_i - \frac{(\Sigma \hat{w}_i d_i)(\Sigma \hat{w}_i z_i)}{\Sigma \hat{w}_i}$$

$$D = \sum \hat{w}_i z_i^2 - \frac{(\Sigma \hat{w}_i z_i)^2}{\Sigma \hat{w}_i}$$

and

$$\chi_1^2 = \frac{N^2}{D}$$

When the cells or classes represent a single x variable, interpretation of the finding of significant variation in δ_i by the preceding methods is straightforward. With two x's, rejection of the null hypothesis does not reveal whether the variation in δ_i is associated primarily with x_1, with x_2, or partly with both. Further, if δ_i varies moderately with one of the x's but not with the other, the F test may lack the power to reject the null hypothesis; however, a test directed at x_1 and x_2 separately may reveal the correct state of affairs.

6.5 y AND x's QUANTITATIVE: ADJUSTMENTS BY REGRESSION (COVARIANCE)

When y and the x's are quantitative, an approach that avoids matching first constructs a mathematical model for the regression of y on the x's, usually assumed of the same form in each population. This regression is then estimated from the sample data and used to adjust $\bar{y}_1 - \bar{y}_2$ in the unmatched samples for differences between the x distributions in the two populations.

In practice, linear regression is the most-frequent form. With k x variables, where x_{j1u} and x_{j2v} denote sample members of the jth x variate in populations 1 and 2, the linear model is

$$y_{1u} = \tau_1 + \sum_{j=1}^{k} \beta_j x_{j1u} + e_{1u}; \quad y_{2v} = \tau_2 + \sum_{j=1}^{k} \beta_j x_{j2v} + e_{2v} \quad (6.5.1)$$

This assumes a constant effect $\delta = \tau_1 - \tau_2$ of the difference between the two treatments. Assuming that e_{1u} and e_{2v} have the same variance, the sums of squares and products $\Sigma(yx_j)$, $\Sigma(x_jx_j)$ and $\Sigma(x_jx_m)$ in the normal equations are the pooled within-treatment values. The notation $\Sigma(yx_j)$ is shorthand; in a notation used earlier it means $(yx_j)_1 + (yx_j)_2$ and each of these terms is a sum of products of deviation scores for the treatment group indicated by the trailing subscript. The adjusted mean difference is

$$\bar{y}_{1a} - \bar{y}_{2a} = \bar{y}_1 - \bar{y}_2 - \sum_{j=1}^{k} b_j(\bar{x}_{j1} - \bar{x}_{j2}) \quad (6.5.2)$$

where the b_j are the estimated regression coefficients. With random samples from each population and a correct mathematical model, the adjusted mean difference is an unbiased estimate of δ under this model.

For the standard error of $\bar{y}_{1a} - \bar{y}_{2a}$ we need s^2, the pooled mean-square deviation from the multiple regression, with $(n_1 + n_2 - k - 2)$ d.f. and the inverse $C = \|c_{jm}\|$ of the matrix $\|\Sigma(x_jx_m)\|$ in the normal equations. The standard error of $\bar{y}_{1a} - \bar{y}_{2a}$ equals

$$s\left(\frac{1}{n_1} + \frac{1}{n_2} + \sum_{j=1}^{k} c_{jj}\bar{d}_j^2 + 2\sum_{j=1}^{k}\sum_{m>j}^{k} c_{jm}\bar{d}_j\bar{d}_m \right)^{1/2} \quad (6.5.3)$$

where $\bar{d}_j = (\bar{x}_{j1} - \bar{x}_{j2})$.

Two precautions are worth noting. Although linear-regression adjustments are the most widely used and are often assumed to hold without checking, we can examine and test for the simpler types of curvature in the regression of y on any x_j by adding a variate $x_{k+1} = x_j^2$ to the model. With large samples and a good computer program, the practice of adding a term in x_j^2 is worthwhile when there are reasons to expect curvature or indications of it. Sometimes a linear regression on a simple transform of x such as $\log x$ or e^{-x} is a satisfactory alternative.

Another precaution in a two-sample regression is to estimate the regression separately in each sample and compare the regression coefficients. The method of adjustment in this section assumes that the regression is the same in the two populations. Possible alternatives when the regression is linear, but differs in the two populations, are outlined in Section 6.7.

Comparison of the regressions in the two samples might reveal that although the regressions appeared to have the same shape, the residual mean squares s_1^2 and s_2^2 are substantially different. In this event, more-precise estimates of the b_j would probably be obtained by weighting the contribution from each sample by $1/s_i^2$ when forming $\Sigma(yx_j)$ and $\Sigma(x_j x_m)$, instead of simply adding. The gain in precision as it affects the estimated treatment effect is, however, usually small.

6.6 REGRESSION ADJUSTMENTS WITH SOME x's CLASSIFIED

The regression method applies most naturally when y and all the x's are quantitative. If one or more of the x's are classified while the others are quantitative, there are two almost-equivalent methods of making the adjustments. To take the simplest case, suppose that one of the x's is a two-class variate. The subscripts $t = 1, 2$ denote the populations or treatments, $i = 1, 2$ the classes, and $j = 1, 2, \ldots, k$ the quantitative x variates. The linear model is assumed to be

$$y_{tiu} = \tau_t + \gamma_i + \sum_{j=1}^{k} \beta_j x_{jtiu} + e_{tiu} \qquad (u = 1, 2, \ldots, n_{ti}) \quad (6.6.1)$$

1. The first method is the *analysis of covariance*. For the quantitative x's, calculate the quantities $\Sigma(yx_j)$ and $\Sigma(x_j x_m)$ from the pooled sums of squares or products within classes and treatments. These will have $(n_1 + n_2 - 4)$ d.f. with two classes and two treatments. Having computed the b_j, take the adjusted y difference, the estimate of $(\tau_1 - \tau_2)$ as

$$\bar{y}_{1a} - \bar{y}_{2a} - \sum_{j=1}^{k} b_j (\bar{x}_{j1a} - \bar{x}_{j2a})$$

Here $\bar{y}_{1a}$, $\bar{y}_{2a}$, $\bar{x}_{j1a}$, and $\bar{x}_{j2a}$ are adjusted means over the two classes, with weights proportional to $n_{1i} n_{2i}/n_i$ in class i.

2. Instead, the adjustments can be performed by an ordinary one-sample multiple regression. The primary advantage of this method is that, at present, computer programs for one-sample multiple regressions are more

widely available than those for the combination of analysis of variance and multiple regression. Construct two dummy x variables: $x_{k+1,\,tiu}$ which has the value $+1$ for all observations from treatment 1 and the value 0 for all observations from treatment 2, and $x_{k+2,\,tiu}$ which has the value $+1$ for all observations in class 1 and the value 0 for all observations in class 2. Fit the multiple-regression model

$$y_{tiu} = \mu + \sum_{j=1}^{k+2} \beta_j x_{jtiu} + e_{tiu} \qquad (6.6.2)$$

Then b_{k+1} is the adjusted estimate of $\tau_1 - \tau_2$. In fact, many computer programs perform the calculations by constructing a third dummy variable, say x_{0tiu}, which takes the value $+1$ for all observations, so that μ in (6.6.2) is replaced by $\beta_0 x_{0tiu}$. It is easily verified that models (6.6.1) and (6.6.2) are identical. The computations in methods 1 and 2, as presented here, are not exactly identical. In method 1, the quantities $\Sigma(yx_j)$ and $\Sigma(x_j x_m)$ for the quantitative x's are based on $(n_1 + n_2 - 4)$ d.f., while in method 2 they are, in effect, based on $(n_1 + n_2 - 3)$ d.f.—the extra d.f. being that for the treatments-by-classes interaction. Any difference in results should be very minor in practice.

With three classes, two dummy x variables are needed for the classification: The first can take the value 1 in class 1 and the value 0 elsewhere; the second takes the value 1 in class 2 and the value 0 elsewhere. Cohen (1968), in describing this technique, has illustrated five equivalent sets of three dummy x variables when there are four classes. Any two sets that are linear transforms of one another are equivalent.

If x is an ordered classification with c classes, a possible alternative is to assign a score $x_{k+1,i}$ to the ith class, creating a single x instead of $(c - 1)$ dummy x's to describe class effects, as suggested by Billewicz (1965). The success of this method depends, of course, on how well the assigned scores are linearly related to y.

Suppose now that there are two classified x's—one with four classes and one with three classes—creating 12 individual cells. The possibilities are to have 11 dummy x's for the effects of the 12 cells, to assume that the effects of the two classifications on y are additive, creating $3 + 2 = 5$ dummy x's for the individual effects of each classification, or, with ordered classifications, to create two sets of scores defining two x variables. Billewicz (1965) reports that the score method did well in removing between-cell bias in a constructed example in which the effects of the two x's were not strictly additive.

6.7   EFFECT OF REGRESSION ADJUSTMENTS ON BIAS IN $\bar{y}_1 - \bar{y}_2$

With y and the x's quantitative, the conditions necessary for fully effective performance of the regression adjustments in removing bias in $\bar{y}_1 - \bar{y}_2$ are (1) the regression of y on the x's is the same in both populations (apart from any difference in the level of the means due to the difference in treatments), (2) the correct mathematical form of the regression has been fitted, and (3) the x's have been measured with negligible error (see Section 6.10).

If these conditions hold, the regression adjustment removes all initial bias. Its performance in this respect is superior to matching and to adjustment by subclassification.

We now consider the failure of condition (1) for linear regressions with different slopes in the two populations. This case was discussed briefly in Section 5.7 with respect to matching, where the conclusion was reached that matching is not appropriate. Regression adjustments are capable of treating this case, but require a judgment as to whether the difference between the regressions in the two populations actually represents confounding effects in treatment. To take the simplest illustration, suppose that the model is

$$y_{1u} = \tau_1 + \beta_1 x_{1u} + e_{1u}; \quad y_{2u} = \tau_2 + \beta_2 x_{2u} + e_{2u} \qquad (6.7.1)$$

It follows that

$$E(\bar{y}_1 - \bar{y}_2) = \tau_1 - \tau_2 + \beta_1 \bar{x}_1 - \beta_2 \bar{x}_2$$

Unbiased estimates of β_1 and β_2 can be obtained and substituted to give an unbiased estimate of $\tau_1 - \tau_2$. The extension to multivariate linear regression is straightforward.

However, an alternative interpretation of (6.7.1), as mentioned previously, is that the effect of the difference in treatments depends on the level of x. Suppose that the regression of y on x is β_2 in each population. Treatment 2 is a control treatment with effect τ_2, while the effect of treatment 1 is $\tau_1 + \delta x$ when applied to a subject whose level is x. The model is then

$$y_{1u} = \tau_1 + (\delta + \beta_2)x + e_{iu}; \quad y_{2u} = \tau_2 + \beta_2 x + e_{2u} \qquad (6.7.2)$$

Belson (1956) has suggested that in making regression adjustments, the estimate b_2 of the regression coefficient from the control sample be used. That is, his adjusted mean difference is

$$\bar{y}_{1a} - \bar{y}_{2a} = (\bar{y}_1 - \bar{y}_2) - b_2(\bar{x}_1 - \bar{x}_2)$$

From (6.7.1) we find

$$E(\bar{y}_{1a} - \bar{y}_{2a}) = \tau_1 - \tau_2 + \delta\bar{x}_1 + \beta_2(\bar{x}_1 - \bar{x}_2) - \beta_2(\bar{x}_1 - \bar{x}_2)$$

$$= \tau_1 - \tau_2 + \delta\bar{x}_1$$

Thus Belson's method estimates the average effect of treatment 1 (as compared with the control) on the persons in sample 1. In some studies this is a quantity of interest to report [see Cochran (1969)]

6.8   EFFECT OF CURVATURE ON LINEAR-REGRESSION ADJUSTMENTS

The effect of linear-regression adjustment on the bias in $\bar{y}_1 - \bar{y}_2$ when the relationship between y and x is monotone and moderately curved has been investigated by Rubin (1973). He dealt with the functions $y = e^{\pm x/2}$ and $y = e^{\pm x}$, with y monotone and quadratic in x, assuming the same form of regression in both populations.

In such cases linear-regression adjustments are still highly effective in removing an initial bias in $\bar{y}_1 - \bar{y}_2$, provided that x has the same variance in the two populations and that the distribution of x is symmetric or nearly symmetric. However, as with matching, the condition $\sigma_{1x}^2 = \sigma_{2x}^2$ is important.

The situation when y is quadratic provides insight on these results. Assume the model

$$y_{tu} = \tau_t + c_1 x_{tu} + c_2 x_{tu}^2 + e_{tu} \tag{6.8.1}$$

where $t = 1, 2$ and $u = 1, 2, \ldots, n$. Following Rubin we consider the bias in $\bar{y}_1 - \bar{y}_2$ conditional on the set of x's that arose in the two samples. For random samples the initial conditional bias is

$$E_c(\bar{y}_1 - \bar{y}_2) - (\tau_1 - \tau_2) = c_1(\bar{x}_1 - \bar{x}_2) + \frac{c_2\Sigma(x_{1u}^2 - x_{2u}^2)}{n}$$

It is convenient to use the notation $s_t^2 = \Sigma(x_{tu} - \bar{x}_t)^2/n$, and $k_{3t} = \Sigma(x_{tu} - \bar{x}_t)^3/n$. Then

$$E_c(\bar{y}_1 - \bar{y}_2) - (\tau_1 - \tau_2) = c_1(\bar{x}_1 - \bar{x}_2) + c_2(\bar{x}_1^2 - \bar{x}_2^2) + c_2(s_1^2 - s_2^2)$$

$$\tag{6.8.2}$$

For the regression adjustment we assume that the pooled within-samples regression coefficient is used, that is,

$$b_p = \frac{\Sigma y_{1u}(x_{1u} - \bar{x}_1) + \Sigma y_{2u}(x_{2u} - \bar{x}_2)}{\Sigma(x_{1u} - \bar{x}_1)^2 + \Sigma(x_{2u} - \bar{x}_2)^2}$$

By substitution for y from the model (6.8.1), b_p is found in large samples to be a consistent estimate of

$$c_1 + 2c_2 \left(\frac{\bar{x}_1 s_1^2 + \bar{x}_2 s_2^2}{s_1^2 + s_2^2} \right) + c_2 \left(\frac{k_{31} + k_{32}}{s_1^2 + s_2^2} \right)$$

Consequently, the remaining conditional bias in $\bar{y}_1 - \bar{y}_2$ after adjustment by $-b_p(\bar{x}_1 - \bar{x}_2)$ approximates

$$-c_2(\bar{x}_1 - \bar{x}_2)^2 \left(\frac{s_1^2 - s_2^2}{s_1^2 + s_2^2} \right) + c_2(s_1^2 - s_2^2) - c_2(\bar{x}_1 - \bar{x}_2) \left(\frac{k_{31} + k_{32}}{s_1^2 + s_2^2} \right)$$

$$(6.8.3)$$

Suppose now that x has differing means but the same variance in the two populations. From (6.8.2) the initial bias approximates

$$c_1(\bar{x}_1 - \bar{x}_2) + c_2(\bar{x}_1^2 - \bar{x}_2^2)$$

From (6.8.3) the bias after adjustment approximates

$$- \frac{c_2(\bar{x}_1 - \bar{x}_2)(k_{31} + k_{32})}{(s_1^2 + s_2^2)}$$

which is small or negligible if the distribution of x is symmetric or nearly symmetric. Thus linear adjustments are highly effective in this case.

In more-general cases with a quadratic regression, comparison of (6.8.2) and (6.8.3) indicates that the average relative sizes of the initial and final biases depend on the sizes and signs of the linear- and quadratic-regression coefficients, c_1 and c_2, on the sizes and signs of $\mu_{1x} - \mu_{2x}$ and $\sigma_{1x}^2 - \sigma_{2x}^2$, and on the amounts of skewness. No simple overall summary statement is possible. With $e^{\pm x/2}$ Rubin found that linear-regression adjustment was successful if $\sigma_{1x}^2 = \sigma_{2x}^2$, but that the adjustment either overcorrected or undercorrected when these variances were unequal, as shown in the first line of Table 6.8.1.

Table 6.8.1. Percent Bias Removed by (1) Linear-Regression (LR) Adjustment on Random Samples, (2) "Nearest Available" Matching, and (3) Linear-Regression Adjustment on Matched Samples [Both Samples of Size 50; $B = \mu_{1x} - \mu_{2x} = \frac{1}{2}\sigma_x$, where $\sigma_x^2 = \frac{1}{2}(\sigma_{1x}^2 + \sigma_{2x}^2)$]

Percent Bias Removed by (1), (2), and (3)	$\sigma_{1x}^2 = \sigma_{2x}^2 = 1$		$\sigma_{1x}^2 = \frac{2}{3}, \sigma_{2x}^2 = \frac{4}{3}$		$\sigma_{1x}^2 = \frac{4}{3}, \sigma_{2x}^2 = \frac{2}{3}$	
	$e^{x/2}$	$e^{-x/2}$	$e^{x/2}$	$e^{-x/2}$	$e^{x/2}$	$e^{-x/2}$
(1) LR[a]	101	101	146	80	80	146
(2) Matching[b]:						
$N/n = 2$	74	94	96	99	45	81
$N/n = 3$	87	98	98	100	60	89
$N/n = 4$	92	99	99	100	65	94
(3) Both[b]:						
$N/n = 2$	102	100	101	100	100	111
$N/n = 3$	100	100	100	100	100	108
$N/n = 4$	100	100	100	100	100	107

[a] With LR on random samples, results are for the pooled within-sample regression; on matched samples, results are for the regression from differences between members of a pair.
[b] N is the size of the reservoir supplying matches to the n members of the target sample.

Table 6.8.1, taken from Rubin (1970), compares "nearest available matching," linear regression applied to random samples, and linear regression applied to the matched samples for $E(y) = e^{\pm x/2}$. The results shown are for a bias in x equal to half the average σ_x—a fairly substantial bias. When $\sigma_{1x}^2 = \sigma_{2x}^2$, linear regression is superior. When $\sigma_{1x}^2 \neq \sigma_{2x}^2$ both methods are erratic and neither method is consistently superior. However, linear regression applied to matched samples was superior to either method and was highly effective. The regression adjustments on matched samples usually performed best when the regression coefficients were estimated from the differences between members of each matched pair. This is the method that would normally be used in matched samples from the viewpoint of analysis of variance.

6.9 EFFECTIVENESS OF REGRESSION ADJUSTMENTS ON PRECISION

As with matching, regression adjustments on random samples may be made in order to increase precision in studies in which the investigator is not concerned with the danger of bias. We assume first a linear regression of y

on a single x, the same in both populations. We suppose that there is an initial bias $B\sigma_x$ in x, since regression adjustments on random samples completely remove this bias. Therefore, there is interest in noting the precision of regression adjustments when $B \neq 0$ as well as when $B = 0$. Using the model, we find that

$$\bar{y}_1 - \bar{y}_2 = \tau_1 - \tau_2 + \beta(\bar{x}_1 - \bar{x}_2) + (\bar{e}_1 - \bar{e}_2) \qquad (6.9.1)$$

the adjusted estimate is

$$\bar{y}_{1a} - \bar{y}_{2a} = \bar{y}_1 - \bar{y}_2 - b(\bar{x}_1 - \bar{x}_2) = (\tau_1 - \tau_2) + (\bar{e}_1 - \bar{e}_2)$$

$$- (b - \beta)(\bar{x}_1 - \bar{x}_2) \qquad (6.9.2)$$

Hence the conditional variance for two samples of size n is

$$V_c(\bar{y}_{1a} - \bar{y}_{2a}) = \frac{2}{n}\sigma_e^2 + \frac{(\bar{x}_1 - \bar{x}_2)^2}{\Sigma_{xx}}\sigma_e^2 \qquad (6.9.3)$$

where Σ_{xx} is the denominator of b, the pooled within-samples $\Sigma(x - \bar{x})^2$. The average value of (6.9.3) in random samples of size n is approximately

$$V(\bar{y}_{1a} - \bar{y}_{2a}) = \frac{2}{n}\sigma_e^2 + \left(B^2 + \frac{2}{n}\right)\frac{\sigma_e^2}{2(n - 2)} \qquad (6.9.4)$$

The expression is correct when x is normal.

In the "no bias" situation ($B = 0$) the leading term in (6.9.4) for large n is $V(\bar{y}_{1a} - \bar{y}_{2a}) = 2\sigma_e^2/n = 2\sigma_y^2(1 - \rho^2)/n$. For within-class matching with the model, $V(\bar{y}_1 - \bar{y}_2)$ is $2\sigma_y^2(1 - f\rho^2)/n$, as given in Section 5.8, where f is the fractional reduction in $V(\bar{x}_1 - \bar{x}_2)$ due to matching. Thus regression adjustments on large random samples give higher precision than within-class matching in the "no bias" case under a linear model. They should perform about as well as tight caliper matching and "nearest available" matching based on a large reservoir for which f is near 1.

When there is initial bias, $B \neq 0$, the leading term in (6.9.4) for large n is

$$V(\bar{y}_{1a} - \bar{y}_{2a}) \doteq 2\sigma_y^2(1 - \rho^2)(1 + \tfrac{1}{4}B^2)/n$$

In this case, regression applied to pair-matched samples would be expected to be more precise than regression on random samples, since pairing reduces $E(\bar{x}_1 - \bar{x}_2)^2$ in (6.9.3).

Under a linear model, the conclusions are little altered when regression adjustments are made on k x variables from random samples. If all x's have the same means in the two populations,

$$V(\bar{y}_{1a} - \bar{y}_{2a}) \doteq \frac{2\sigma_y^2}{n}(1 - R^2)\left(1 + \frac{k}{2n - k - 3}\right) \qquad (6.9.5)$$

where R^2 is the squared multiple correlation coefficient between y and the x's. If $k/2n$ is negligible, this variance is practically $2\sigma_y^2(1 - R^2)/n$. Within-class matching with the same number of classes per variable gives $2\sigma_y^2(1 - fR^2)/n$, which is a larger value.

When there is initial bias, (6.9.5) also contains a quadratic expression in the biases B_j of the variables x_j; this term is of the same order as $2\sigma_y^2(1 - R^2)/n$. As before, regression applied to pair-matched samples should perform better than regression on random samples.

Using experimental sampling on a computer, Billewicz (1965) made comparisons of the precisions of within-class matching, regression on random samples, and regression applied to matched samples under a variety of situations. He uses two groups—treated and control. His results, reported here, concern relative precision in the "no bias" situation.

1.   For a linear-regression model with y and x quantitative, regression was more precise than within-class matching with three or four groups, by amounts that agreed well with those given here.

2.   Billewicz also made this comparison, with $n = 40$ in each sample, for three different nonlinear regression models, $y = 0.4x - 0.1x^2$, $y = 0.8x - 0.14x^2$, and $y = \tanh x$, with x following $N(0, 1)$ in both populations. These amounts of nonlinearity were detectable in 12.3, 20.3, and 19.8% of his samples. Despite the use of an incorrect model, *linear*-regression adjustments were superior in precision to matching with three or four classes.

3.   When linear regressions have different slopes in the two populations, Billewicz indicates the importance of detecting this situation and the difficulties of interpretation to which we have referred. Matched pairs, regression analysis of random samples, and regression analysis applied to frequency matched samples were about equally effective in detecting the difference in slopes. He concludes that the average user of matched samples would be unlikely to examine his sample in this respect, since the concept of matching is directed toward finding a single overall effect of treatment.

6.10   EFFECT OF ERRORS IN THE MEASUREMENT OF X

Sometimes confounded x variables are difficult to measure and hence are measured with substantial errors. In large-sample studies a crude measuring device may be used for reasons of expense or because accurate measurement requires trained personnel who are in short supply. As noted by Lord (1960) and other investigators, regression adjustments fail to remove all the initial bias when the x's are measured with error. Their effectiveness in increasing precision is also reduced.

The symbol x denotes the fallible measurement actually made, while X denotes the correct value, and e, the error. The simplest model with two populations is

$$y_{1u} = \tau_1 + \beta X_{1u} + e_{1u}; \qquad y_{2u} = \tau_2 + \beta X_{2u} + e_{2u}$$

$$x_{1u} = X_{1u} + h_{1u}; \qquad x_{2u} = X_{2u} + h_{2u}$$

where h_{1u} and h_{2u} are the errors of measurement. The errors h are assumed independent of e, but h_{tu} and X_{tu} may be correlated.

Lindley (1947) has shown that even if h and X are independent, the regression of y on the fallible x is not linear unless the distributions of h and X belong, in a certain sense, to the same type (e.g., both χ^2 or normal). However, there is some evidence (Cochran, 1970a) that the linear component is dominating, and in this discussion, nonlinearity will be ignored. The slope β' of the linear component is

$$\beta' = \beta \left(\sigma_X^2 + \sigma_{Xh}\right) / \left(\sigma_X^2 + 2\sigma_{Xh} + \sigma_h^2\right) \qquad (6.10.1)$$

here σ_{Xh} is the population covariance of X and h. If b' is the estimated regression coefficient of y on x, then for given $\bar{x}_1 - \bar{x}_2$,

$$E_c\left(\bar{y}_{1a} - \bar{y}_{2a}\right) = E\left(\bar{y}_1 - \bar{y}_2\right) - \beta'\left(\bar{x}_1 - \bar{x}_2\right)$$

$$= \tau_1 - \tau_2 + \left(\beta - \beta'\right)\left(\bar{x}_1 - \bar{x}_2\right)$$

Thus, conditionally, a fraction $(\beta - \beta')/\beta$ of the initial bias remains after adjustment.

If h and X are uncorrelated, $\beta' = \beta\sigma_X^2/\sigma_x^2 = G\beta$, where G is a quantity often called the "reliability" of the measurement. Thus $100G$ is the percentage of the initial bias that is removed and $100(1 - G)$ is the percentage remaining. One method attempts to remove all the initial bias by regression adjustments. It estimates σ_h^2 and σ_X^2 and hence G by an auxiliary study, and thus obtains a consistent estimate of β which is used instead of b' in making the regression adjustment. Lord (1960) addresses this problem.

Such errors of measurement also affect the performance of within-class matching and adjustment by weighted means when a fallible quantitative x is replaced by a classification in order to use these methods. These methods still produce a fractional reduction of amount f in the initial bias of $\bar{x}_1 - \bar{x}_2$, as discussed in Sections 5.4 and 6.2, but because of the errors of measurement, this creates a fractional reduction of only fG in the initial bias of $\bar{X}_1 - \bar{X}_2$ and hence of $\bar{y}_1 - \bar{y}_2$ under the linear-regression model. The relative performance of regression on random samples and within-class matching is, therefore, unaffected by such errors of measurement in x.

The gain in precision due to regression adjustments in the "no bias" case is also affected by errors of measurement in x. With h and X uncorrelated, the population correlation ρ' between y and x is

$$\rho' = \frac{\sigma_{yx}}{\sigma_y \sigma_x} = \frac{\beta(\sigma_X^2 + \sigma_{Xh})}{\sigma_y \sigma_x} = \frac{\rho \sigma_X}{\sigma_x} = \rho \sqrt{G}$$

Hence the residual variance from the regression is $\sigma_y^2(1 - G\rho^2)$, instead of $\sigma_y^2(1 - \rho^2)$. For a given reliability of measurement G, the relative loss of precision is greatest when ρ^2 is high, that is, when X is a very good predictor of y.

6.11   MATCHING AND ADJUSTMENT COMPARED: IN EXPERIMENTS

We start with $2n$ subjects, presumed drawn at random from the sampled population. In this population the linear-regression model is

$$y = \alpha + \beta(x - \mu) + e$$

where the residual e is assumed to have mean 0 and variance σ_e^2 for any fixed x. If n subjects are assigned at random to each of two treatments, T_1 and T_2,

$$\bar{y}_1 - \bar{y}_2 = \tau_1 - \tau_2 + \beta(\bar{x}_1 - \bar{x}_2) + \bar{e}_1 - \bar{e}_2$$

Hence

$$E(\bar{y}_1 - \bar{y}_2) = \tau_1 - \tau_2 \quad \text{(no bias)}$$

and

$$V(\bar{y}_1 - \bar{y}_2) = \frac{2}{n}(\beta^2 \sigma_x^2 + \sigma_e^2) = \frac{2}{n}\left[\rho^2 \sigma_y^2 + (1 - \rho^2)\sigma_y^2\right]$$

since $\beta \sigma_x = \rho \sigma_y$. The penalty for failure to control x is loss of precision, since $V(\bar{y}_1 - \bar{y}_2)$ is inflated by the term $2\beta^2\sigma_x^2/n$, or by a factor of $1/(1 - \rho^2)$.

Matching on x

The $2n$ values of x are ranked in decreasing order. Of the two highest x's, one x is assigned at random to T_1 and the other x to T_2, and so forth for succeeding pairs. This gives

$$(\bar{y}_1 - \bar{y}_2)_m = \tau_1 - \tau_2 + \beta(\bar{x}_1 - \bar{x}_2)_m + \bar{e}_1 - \bar{e}_2.$$

The quantity $(\bar{x}_1 - \bar{x}_2)_m$ will not be exactly zero in this method of matching, but will have a variance that can be calculated from the variances and covariances of the order statistics. For n exceeding 50 it appears that this variance is negligible with x approximately normal, so

$$V(\bar{y}_1 - \bar{y}_2)_m = \frac{2}{n}\sigma_y^2(1 - \rho^2)$$

Regression Adjustment

When matching is not used, each treatment is assigned at random to n subjects. The pooled sample estimate b of β is

$$b = \frac{\Sigma_1 y(x - \bar{x}_1) + \Sigma_2 y(x - \bar{x}_2)}{\Sigma_1(x - \bar{x}_1)^2 + \Sigma_2(x - \bar{x}_2)^2} = \beta + \frac{\Sigma_1 e(x - \bar{x}_1) + \Sigma_2 e(x - \bar{x}_2)}{\Sigma_1(x - \bar{x}_1)^2 + \Sigma_2(x - \bar{x}_2)^2}$$

as is found when we substitute $y = \alpha + \beta(x - \mu_x) + e$ in the formula for b. For fixed x's the quantity $b - \beta$ is a random variable $\bar{e}_w$ in the e's with mean 0 and variance $\sigma_e^2/(\Sigma_1 + \Sigma_2)$, where $\Sigma_1 = \Sigma_1(x - \bar{x}_1)^2$, and so forth.
   Hence the adjusted estimate

$$(\bar{y}_1 - \bar{y}_2) - b(\bar{x}_1 - \bar{x}_2) = \tau_1 - \tau_2 + (\beta - b)(\bar{x}_1 - \bar{x}_2) + \bar{e}_1 - \bar{e}_2$$

$$= \tau_1 - \tau_2 + (\bar{e}_1 - \bar{e}_2) - (\bar{x}_1 - \bar{x}_2)\bar{e}_w$$

The variance of the regression-adjusted estimate is, therefore, for fixed x's because $\bar{e}_w$ is uncorrelated with $\bar{e}_1$ and $\bar{e}_2$,

$$\frac{2}{n}\sigma_e^2 + \frac{(\bar{x}_1 - \bar{x}_2)^2\sigma_e^2}{\Sigma_1 + \Sigma_2} = \frac{2}{n}\sigma_y^2(1 - \rho^2)\left(1 + \frac{n(\bar{x}_1 - \bar{x}_2)^2}{2(\Sigma_1 + \Sigma_2)}\right)$$

For x normal the second term in the large parentheses may be shown to have mean $1/2(n - 2)$, which is only about 0.01 when n is 50.

Thus in experiments with $n = 50$ or more and a linear model, matching and regression adjustment are about equally effective. Their purpose is to increase the precision of the estimate of $\tau_1 - \tau_2$ and their effect is to reduce the term σ_y^2 in $V(\hat{\tau}_1 - \hat{\tau}_2)$ to $\sigma_y^2(1 - \rho^2)$.

6.12   MATCHING AND ADJUSTMENT COMPARED: IN OBSERVATIONAL STUDIES

In an observational comparison of two treatments, the investigator begins with *two* populations—one for each treatment. The investigator has chosen to study these two populations but did not create them. The investigator must suppose that, in general, the two populations will have different means (μ_{1y}, μ_{2y}) and (μ_{1x}, μ_{2x}). In their simplest form the regression models in the two populations become, for subject j in sample 1 and subject k in sample 2,

$$y_{1j} = \mu_{1y} + \tau_1 + \beta(x_{1j} - \mu_{1x}) + e_{1j} \qquad (6.12.1)$$

and

$$y_{2k} = \mu_{2y} + \tau_2 + \beta(x_{2k} - \mu_{2x}) + e_{2k} \qquad (6.12.2)$$

(For this illustration it is assumed that uncontrolled variables whose effects on y are summed in the terms e_{1j} and e_{2j} behave as random variables.)

Suppose first that random samples are drawn from the respective sampled populations making no attempt to control for x. Then

$$\bar{y}_1 = \mu_{1y} + \tau_1 + \beta(\bar{x}_1 - \mu_{1x}) + \bar{e}_1 \qquad (6.12.3)$$

and

$$\bar{y}_2 = \mu_{2y} + \tau_2 + \beta(\bar{x}_2 - \mu_{2x}) + \bar{e}_2 \qquad (6.12.4)$$

In repeated sampling, $E(\bar{x}_i) = \mu_{ix}$ and $E(\bar{e}_i) = 0$ $(i = 1, 2)$. Hence

$$E(\bar{y}_1 - \bar{y}_2) = \tau_1 - \tau_2 + (\mu_{1y} - \mu_{2y})$$

The estimate $\bar{y}_1 - \bar{y}_2$ is now biased by the amount $\mu_{1y} - \mu_{2y}$, with the bias favoring the treatment given to the population with the *higher* mean of y. As Campbell and Erlebacher (1970) and Campbell and Boruch (1975) have stressed, this bias produces an *underestimate* in the beneficial effect of a program given to a sample $\bar{y}_2$ from a disadvantaged population.

With random samples from the two populations the standard error (SE) of $\bar{y}_1 - \bar{y}_2$ is $\sqrt{\sigma_{1y}^2 + \sigma_{2y}^2} / \sqrt{n}$.

The ratio of the bias $\mu_{1y} - \mu_{2y}$ to this SE is

$$\frac{\sqrt{n}\,(\mu_{1y} - \mu_{2y})}{\sqrt{\sigma_{1y}^2 + \sigma_{2y}^2}}$$

This increases indefinitely as n grows. Tests of significance of the null hypothesis (NH) $\tau_1 - \tau_2$ are likely to reject the NH even when it is actually true, so that no clear interpretation can be given to rejection of the NH by the test. The interpretation of a nonsignificant result is also obscured by the possibility that $\tau_1 - \tau_2$ and $\mu_{1y} - \mu_{2y}$ have similar magnitudes and opposite signs. The relative sizes of the bias $\mu_{1y} - \mu_{2y}$ to the true treatment difference $\tau_1 - \tau_2$ obviously affects any conclusions drawn about the relative merits of the treatments.

Thus in observational studies, matching and regression have two objectives: to remove or reduce bias and to increase precision by reducing the SE of $\bar{y}_1 - \bar{y}_2$. Of these, it is reasonable to regard reduction of bias as the more-important objective. A highly precise estimate of the wrong quantity is of limited use.

Under this model of parallel linear regressions, the complete removal of bias by either matching or regression adjustment requires that the following condition hold:

$$\beta = (\mu_{1y} - \mu_{2y})/(\mu_{1x} - \mu_{2x}) \tag{6.12.5}$$

This condition, in turn, is equivalent to each of the following:

1. Both populations (in the absence of treatment) have the same regression line.
2. The regression of y on x within the populations is equal to the regression between populations.

If (6.12.5) does not hold, the bias in estimating $\tau_1 - \tau_2$ is, after adjustment, equal to

$$(\mu_{1y} - \mu_{2y}) - \beta(\mu_{1x} - \mu_{2x})$$

(The appendix to this section gives the justification for these statements.)

Without evidence that the regression lines are the same in the two populations, the attitude of the investigator may have to be that matching and regression adjustment leave some residual bias. The investigator hopes that this bias is only a small fraction of the original bias $\mu_{1y} - \mu_{2y}$ —sufficiently small in relation to $\tau_1 - \tau_2$ so that conclusions drawn about the treatments are little affected.

As Bartlett (1936) and Lord (1960) have stated, both matching and regression in this situation involve an element of unverifiable extrapolation. To take an extreme case, suppose $\mu_{1x} < \mu_{2x}$, the difference being so large that no member of sample 1 has a value as high as $\bar{x}_2$. We can still apply the regression adjustment. Formally, this adjusts $\bar{y}$ to its predicted value when the mean of the accompanying x's is $\bar{x}_2$, so that the adjusted $\bar{y}_1$ becomes comparable with $\bar{y}_2$. But this adjusted value is purely hypothetical when we have no y_{1j} value with an accompanying x as high as $\bar{x}_2$. In less-extreme cases the extrapolation is more moderate.

APPENDIX TO SECTION 6.12

Matching on x

We try to find matched pairs of subjects from the two populations such that $x_{1j} - x_{2j}$ is small in the jth pair ($j = 1, 2, \ldots, n$). Incidentally, if μ_{1x} and μ_{2x} differ substantially, matching is often a slow process, requiring large reservoirs of subjects. This may require, for instance, finding subjects with unusually high x's from population 1 to pair with low x's from population 2.

Successful matching will make $\bar{x}_1 - \bar{x}_2$ negligible. In this event, from (6.12.3) and (6.12.4),

$$E(\bar{y}_1 - \bar{y}_2)_m = \tau_1 - \tau_2 + \mu_{1y} - \mu_{2y} - \beta(\mu_{1x} - \mu_{2x})$$

Hence, all the bias is removed by matching if

$$\mu_{1y} - \mu_{2y} = \beta(\mu_{1x} - \mu_{2x}) \tag{6.12.5}$$

This condition can be described in two equivalent ways:

1. From (6.12.1) the regression lines in the two populations may be written (in the absence of any treatment effect)

$$E(y_{1j}|x_{1j}) = \mu_{1y} - \beta\mu_{1x} + \beta x_{1j}$$

and

$$E(y_{2j}|x_{2j}) = \mu_{2y} - \beta\mu_{2x} + \beta x_{2j}$$

The condition $\mu_{1y} - \mu_{2y} = \beta(\mu_{1x} - \mu_{2x})$ then means that these two lines have the same intercepts and slopes, that is, they are identical.

From the results of the study we can test whether the slopes are the same. If they are $y_{1j} - y_{2j}$ should have no regression on x_{1j}. Since x_{1j} and x_{2j} often differ slightly in matched pairs, an approximation is to compute and test the regression of $y_{1j} - y_{2j}$ on $(x_{1j} + x_{2j})/2$. But given only y_{ij} (after treatment) and x_{ij}, we cannot check from the data whether the intercepts would be identical in the absence of treatment effects. If we fit separate parallel lines to the samples from the two populations, the intercepts on the fitted lines will be estimates of

$$\mu_{1y} - \beta\mu_{1x} + \tau_1; \qquad \mu_{2y} - \beta\mu_{2x} + \tau_2$$

They will thus differ by an estimate of the treatment difference $\tau_1 - \tau_2$, if condition (6.12.5) holds.

In the type of study called the *pretest–posttest study*, y is measured both before treatments are applied as well as after a period of application. With such data, coincidence of the regression lines in the absence of treatment can be tested from the pretest data.

2.   The condition for the removal of bias

$$\beta = (\mu_{1y} - \mu_{2y})/(\mu_{1x} - \mu_{2x})$$

can also be described as meaning that the between-population regression of y on x must equal the within-population regression. If we were given the pairs of means μ_{iy} and μ_{ix} for a number of populations, the regression of μ_{iy} on μ_{ix} might appropriately be called the "between population" regression of y on x. With only two populations the slope of this regression is $(\mu_{1y} - \mu_{2y})/(\mu_{1x} - \mu_{2x})$.

Regression Adjustment

Here we assume random samples from the two populations, with no attempt at matching. The adjusted estimate of $\tau_1 - \tau_2$ is

$$(\bar{y}_1 - \bar{y}_2)_{\text{adj}} = \bar{y}_1 - \bar{y}_2 - b(\bar{x}_1 - \bar{x}_2)$$

With random samples and the linear model, $E(b) = \beta$ for any set of x. Further, $E(\bar{x}_1) = \mu_{1x}$ and $E(\bar{x}_2) = \mu_{2x}$. Hence

$$E(\bar{y}_1 - \bar{y}_2)_{\text{adj}} = \tau_1 - \tau_2 + \mu_{1y} - \mu_{2y} - \beta(\mu_{1x} - \mu_{2x})$$

The expression $\mu_{1y} - \mu_{2y} - \beta(\mu_{1x} - \mu_{2x})$ is the bias that remains after regression adjustment; it vanishes when condition (6.12.5) holds. Thus under the linear model both the residual bias and the condition for its complete removal are the same for adjustment by linear regression as for matching.

6.13   A PRELIMINARY TEST OF COMPARABILITY

In deciding whether to match or adjust for an x variable, it has been recommended that consideration be given first to x's in which it is suspected that there will be a bias arising from a difference $\mu_{1x} - \mu_{2x}$ in the means of x. If uncertain whether there is a danger of bias, we might first make a t test of significance of $\bar{x}_1 - \bar{x}_2$ from two random samples of size n. If t is nonsignificant, we judge that the risk of major bias is small and decide not to match or adjust for this x. Tests of significance are often employed as decision rules in this way.

This procedure has been examined [Cochran (1970b)] assuming a linear regression of y on x. If t is significant at some chosen level, a linear-regression adjustment on random samples is made. If t is not significant, the unadjusted estimate $\bar{y}_1 - \bar{y}_2$ is used.

Under the standard linear-regression model the conditional mean of the adjusted y difference given $\bar{x}_1$ and $\bar{x}_2$ is

$$E_c(\bar{y}_{1a} - \bar{y}_{2a}) = \tau_1 - \tau_2 - (\bar{x}_1 - \bar{x}_2)E_c(b - \beta)$$

Now $b - \beta = \Sigma e(x - \bar{x})/\Sigma(x - \bar{x})^2$, where the Σ's are the pooled within-sample sums of squares or products. Consider samples selected so that $t = \sqrt{n}\,|\bar{x}_1 - \bar{x}_2|/\sqrt{2}\,s_x$ is significant. Since this selection is based solely on the values of x and since e and x are independent, $E_c(b - \beta) = 0$ in samples selected in this way. (The conditional variance of b is affected, but not the conditional mean). It follows that the adjusted $\bar{y}_{1a} - \bar{y}_{2a}$ is free from bias when t is significant.

The remaining bias from this process is, therefore,

$$P(|t| < t_0)E_c(\bar{y}_1 - \bar{y}_2) = P(|t| < t_0)\beta E_c(\bar{x}_1 - \bar{x}_2)$$

where t_0 is the critical value of t and the conditional mean is for $|t| < t_0$. The intuitive idea behind the method is, of course, that if $\mu_{1x} - \mu_{2x}$ is large there should be little remaining bias because t is almost certain to be significant. If $\mu_{1x} - \mu_{2x}$ is small, t may be nonsignificant frequently, but the final bias should be small because the initial bias is small.

At some intermediate point we obtain the maximum final bias. Since $\beta(\bar{x}_1 - \bar{x}_2) = \beta\sqrt{2}\,s_x t/\sqrt{n}$ the final bias is of order $1/\sqrt{n}$.

The maximum final bias occurs when the probability of a nonsignificant t is around 0.70 for 5% tests, 0.65 for 10% tests, and 0.60 for 20% tests. Expressed for convenience as a fraction f of the quantity $\sqrt{2}\,\beta\sigma_x/\sqrt{n}$, the values of f vary between 0.72 (20 d.f. for t tests) and 0.66 (∞ d.f.) for 5% tests, 0.49 and 0.45 for 10% tests, and 0.26 and 0.25 for 20% tests. As expected, a larger, that is, less stringent, significance level of t gives a smaller final bias at the expense of more-frequent adjustments.

Is the procedure adequate? Suppose an investigator uses standard elementary formulas for tests of significance of $\bar{y}_1 - \bar{y}_2$ or confidence levels of $\mu_{1y} - \mu_{2y}$ after using this test. That is, the investigator assigns to $\bar{y}_1 - \bar{y}_2$ a standard error $\sqrt{2}\,s_y/\sqrt{n}$, if t is nonsignificant, and to $(\bar{y}_{1a} - \bar{y}_{2a})$ a standard error

$$\sqrt{2}\,s_{y\cdot x}\frac{\left[1 + (\bar{x}_1 - \bar{x}_2)^2/\Sigma\right]^{1/2}}{\sqrt{n}}$$

if t is significant. (In the preceding expression, $s_{y\cdot x}$ is the root of the residual variance about the regression line based on pooled within-group sums of squares and cross-products. Also Σ is the pooled within-group sum of squares of x.) Even with 5% tests, it is found that type-I errors and confidence probabilities are only slightly disturbed.

Alternatively, we might ask whether the maximum remaining bias is negligible with respect to the size of difference δ_y that we are trying to measure; or in other words, whether the ratio $\sqrt{2}\,f\rho\sigma_y/\sqrt{n}\,\delta_y$ is negligible. The answer here is less certain, since it depends on n, ρ, and the ratio δ_y/σ_y. For instance, in some applications an improvement of a new method of treatment over a standard method might be important in practice if $\delta_y/\sigma_y = 0.2$. Taking the maximum f as about 0.7 for 5% tests and $\rho = 0.4$, the ratio of the maximum bias to this δ_y is $\sqrt{2}\,(0.7)(0.4)/0.2\sqrt{n} = 1.98/\sqrt{n}$. If we want the ratio to be less than 10%, we need $n = (19.8)^2 = 392$ in each sample. Unless we have samples at least this large, the ratio will not be negligible (less than 10%).

6.14 SUMMARY

In comparing the means $\bar{y}_1 - \bar{y}_2$ or proportions $\hat{p}_1 - \hat{p}_2$ from two populations, an alternative to matching is to draw random samples from the two populations and make adjustments in the statistical analysis to $\bar{y}_1 - \bar{y}_2$ or $\hat{p}_1 - \hat{p}_2$ in order to reduce bias or increase precision. The method of adjustment depends on the scales in which the variables are measured.

If the y's are quantitative and the x's are classified (or have been made classified), let $\bar{y}_{1i}$ and $\bar{y}_{2i}$ be the sample means of y in the ith cell of this classification. If the effect $\tau_1 - \tau_2$ of the difference in treatments is the same in every cell, any weighted mean $\Sigma w_i(\bar{y}_{1i} - \bar{y}_{2i}) = \Sigma w_i d_i$, with $\Sigma w_i = 1$, controls bias to precisely the same extent as does within-class matching, so that there is little difference between the methods in this respect. The choices of weights that minimize the variance of $\Sigma w_i d_i$ are given in Section 6.2. In particular, optimum weights are proportional to $n_{1i} n_{2i}/(n_{1i} + n_{2i})$ if the within-cell and treatment variances are constant. Under this assumption, two matched samples of size n give a smaller variance than two weighted random samples of size n, but the difference is likely to be minor in the "no bias" case in which the comparision is of most interest.

If y is a $(0, 1)$ variate, many workers have assumed a model in which the effect of the difference between treatments is constant from cell to cell on the scale of logit $p_{ti} = \log(p_{ti}/q_{ti})$. Under this model it may still be desirable to estimate and test a weighted mean difference of the form $\Sigma w_i(\hat{p}_{1i} - \hat{p}_{2i})/\Sigma w_i$. For this purpose, a good choice for testing significance is $v_i = n_{1i} n_{2i}/(n_{1i} + n_{2i})$.

If the treatment effect $\delta_i = \tau_{1i} - \tau_{2i}$ differs from cell to cell, the choice of weights determines the quantity $\Sigma w_i \delta_i$ that is being estimated. In the analysis, possibilities are (1) to use weights derived from a target population that is of interest; (2) to note that the values of $\Sigma w_i \delta_i$ agree well enough for different weighting systems so that the same conclusion or action is suggested; and (3) to decide against estimation of an overall mean and to summarize instead the way in which δ_i varies from cell to cell. A method of testing whether δ_i varies from cell to cell is given, but the interpretation of the test is simple only when a single x variable is involved.

When y and the x's are all quantitative, adjustments for bias may be made on random samples by means of the regression of y on the x's. This method can also include a classified x by the creation of dummy variables to represent class effects or (with ordered classifications) by assigning scores to the classes. In practice, a linear regression with the same slopes in both populations is most commonly assumed, but the method provides tests for differences in slopes and for nonlinearity which help to make the assumed model more nearly correct. Linear-regression adjustments can be used when there are differences in slopes, if these differences are due to the confounding x variables. However, another possible interpretation is that the differences may represent a relation between the effects of the treatments and the level of x.

With regard to the control of bias, the regression method removes all the initial bias, provided that the fitted model is correct in form and the x's are not subject to errors of measurement. In this situation, regression is superior to pair or within-class matching and to adjustment by weighted class means. If adjustment by linear regression is used when the true regression of y on x is monotone and moderately curved [e.g., a quadratic or $E(y) = e^{\pm x/2}$], the available evidence suggests that linear adjustment still removes almost all the bias, provided that $\sigma_{1x} = \sigma_{2x}$ and that the distribution of x is symmetrical. If $\sigma_{1x} \neq \sigma_{2x}$ the performance of linear-regression adjustments on $e^{\pm x/2}$ is erratic. However, linear-regression adjustments on matched samples were highly successful in this situation.

With regard to the precision of $\bar{y}_1 - \bar{y}_2$ in the "no-bias" situation, linear-regression adjustments on random samples were superior under a linear-regression model to within-class matching and almost as good as mean matching and tight caliper matching. With three monotone nonlinear population regressions (two quadratic and one $y = \tanh x$) Billewicz found linear-regression adjustments superior in precision to within-class matching with three or four classes.

By way of an overall comparison, the comparisons made indicate that, with y and the x's quantitative, regression adjustments based on random samples should be superior to within-class matching and probably also superior to a fairly tight caliper matching and "nearest available" pair matching based on a large reservoir, provided that care is taken to fit approximately the correct shape of regression. Even if linear adjustments are routinely applied, they appear to perform about as well as "nearest available" pair matching in the presence of monotone curved regressions. In such cases, however, linear-regression adjustments applied to pair-matched samples are consistently better in removing bias.

If the true regression of y on X is linear, but the measured x is subject to an independent error of measurement, the percentage of bias removed by regression adjustment is reduced, dropping to $100G$, where $G = \sigma_X^2/\sigma_x^2$. The performance of within-class matching is affected similarly.

Finally, one possibility is to adjust for the regression on x as a precaution against bias only if $\bar{x}_1 - \bar{x}_2$ is statistically significant. Under a linear model, this decision rule operates well enough so that type-I errors and confidence probabilities relating to $\bar{y}_1 - \bar{y}_2$ calculated by standard techniques are not much affected.

REFERENCES

Bartlett, M. S. (1936). A note on the analysis of covariance. *J. Agric. Sci.*, **26**, 488–491.

Belson, W. A. (1956). A technique for studying the effects of a television broadcast. *Appl. Statist.*, **5**, 195–202.

Billewicz, W. Z. (1965). The efficiency of matched samples: an empirical investigation. *Biometrics*, **21**, 623–644.

Campbell, D. T. and R. F. Boruch (1975). Making the Case for Randomized Assignment to Treatments by Considering the Alternatives: Six Ways in which Quasi-Experimental Evaluations in Compensatory Education Tend to Underestimate Effects, in C. A. Bennett and A. A. Lumsdaine, Eds. *Evaluation and Experience: Some Critical Issues in Assessing Social Programs.* Academic Press, New York.

Campbell, D. T. and A. E. Erlebacher (1970). How Regression Artifacts in Quasi-Experimental Evaluations Can Mistakenly Make Compensatory Education Look Harmful, in J. Hellmuth, Ed. *Compensatory Education: A National Debate, 3, Disadvantaged Child.* Brunner/Mazel, New York.

Cochran, W. G. (1954). Some methods for strengthening the common χ^2 tests. *Biometrics*, **10**, 417–451 [Collected Works #59].

Cochran, W. G. (1969). The use of covariance in observational studies. *Appl. Statist.*, **18**, 270–275 [Collected Works #92].

Cochran, W. G. (1970a). Some effects of errors of measurement on linear regression. *Proceedings of the 6th Berkeley Symposium*, Vol. 1, University of California Press, pp. 527–539 [Collected Works #95].

Cochran, W. G. (1970b). *Performance of a Preliminary Test of Comparability in Observational Studies* ONR Technical Report No. 29, Department of Statistics, Harvard University, Cambridge, Mass. [Collected Works #96].

Cohen, J. (1968). Multiple regression as a general data-analytic system. *Psychol. Bull.*, **70**, 426–443.

Keyfitz, N. (1966). Sampling variance of standardized mortality rates. *Human Biology*, **38**, 309–317.

Lindley, D. V. (1947). Regression lines and the linear functional relationship. *J. Roy. Statist. Soc. B*, **9**, 218–244.

Lord, F. (1960). Large-sample covariance analysis when the control variable is fallible. *J. Am. Statist. Assoc.*, **55**, 307–321.

Mantel, N. and W. Haenszel (1959). Statistical aspects of the analysis of data from retrospective studies of disease. *J. Natl. Cancer Inst.*, **22**, 719–748.

Meier, P. (1953). Variance of a weighted mean. *Biometrics*, **9**, 59–73.

Rubin, D. B. (1970). The Use of Matched Sampling and Regression Adjustment in Observational Studies. *Ph.D. Thesis, Harvard University, Cambridge, Mass.*

Rubin, D. B. (1973). The use of matched sampling and regression adjustment to remove bias in observational studies. *Biometrics*, **29**, 185–203.

CHAPTER 7

Simple Study Structures

7.1 INTRODUCTION

This chapter discusses the structure of some of the simplest observational-study plans that have been used in the medical and social sciences, when the objective is to measure the effect of a given treatment. A more-extensive catalog of such plans appears in the monograph by Campbell and Stanley (1963/1966), whose primary purpose was to appraise the strengths and weaknesses of such plans in educational research.

7.2 THE SINGLE GROUP: MEASURED AFTER TREATMENT ONLY

The situation in which this is the only kind of data can arise if all persons available for study were exposed to the treatment, if no unexposed group at all comparable can be found, and if no response measurements previous to the application of treatment exist. With this plan there is no basis for a judgment about the effect of treatment, unless we can guess accurately enough what would have happened in the absence of treatment. This could be so, for instance, with an unusual natural calamity. An estimate of the number of deaths caused by an earthquake, based on a count of dead bodies, might be highly accurate, the principal error being the status of missing persons not accounted for, where the number expected to die during the time in question in the absence of an earthquake may be safely regarded as minor.

In the absence of any external comparison group, it may still be possible to learn something if (1) different persons were exposed to the treatment in different degrees, and if (2) it is possible to assign some score or measure to each person representing degree of exposure. This was roughly the situation in the important studies of the aftereffects of the atomic bomb on survivors in Hiroshima. By combining a person's memory of location (distance from the epicenter) and of local shielding by buildings, the survivors could be divided into four exposure groups. Subsequent four-group morbidity and mortality is, of course, no longer a "single-group" study. The group furthest from the epicenter could, in fact, be regarded as scarcely exposed at all to unusual radiation. The main difficulty with this group was that ethnically and socially they appeared somewhat different from the other groups.

In fact, any method that occurs to me for attempting an inference about the effect of the treatment amounts to changing this plan into a different one, either by obtaining comparable measurements before the treatment was applied or by guessing or finding comparison groups exposed to different levels of treatment.

7.3   THE SINGLE GROUP: MEASURED BEFORE AND AFTER TREATMENT

This situation is probably more common than the preceding one as a device from which conclusions about the causal effects of a treatment are attempted. Some new program or law intended to be beneficial in certain respects, for example, fluoridation of water or a change in working conditions, is such that all people in a given community or agency are exposed. In an attempt to evaluate the effects of the program, relevant response variables are recorded either for the community or agency as a whole or for a sample of people, both before the introduction of the treatment and at a time afterward when the treatment is expected to have produced its major effects. It is important that the sample be a random sample and that strong efforts be made to keep nonresponse to a minimum. If the treatment is quick-acting we may plan to use the same sample of people before and after; if not, the samples may be drawn independently or with perhaps some matching, as described in Chapter 5, to secure greater comparability.

With this plan we at least obtain an estimate $\bar{y}_a - \bar{y}_b$ of the time change that took place in a response variable and can attach a standard error to $\bar{y}_a - \bar{y}_b$. There remains the problem of judging how much this change was due to the treatment or to other contributing causes.

With only two sets of observations—before and after—a basic difficulty is that even if $\bar{y}_a - \bar{y}_b$ is clearly statistically significant by the appropriate t test, this difference may be merely the size that commonly occurs due to the multiplicity of other causes that create time changes in the time interval τ involved. Thus even if the investigator can think of no specific alternative source that might provide a rival hypothesis as to the reason for the $\bar{y}_a - \bar{y}_b$

difference, this provides only a subjective basis for a claim that the difference was caused by the treatment. There must be some objective support of the judgment that $\bar{y}_a - \bar{y}_b$ is larger than commonly occurs in a time interval of this size. Seeking evidence for this judgment usually involves getting data from populations not subject to the treatment during this time interval; in other words, changing from a single-group to at least a two-group comparison. Data from such other populations is often helpful in judgment even if they cannot be regarded as strictly comparable. With a single group or population it helps also if a series of observations at intervals τ have been made both before and after treatment, since we get some information on the sizes of changes in $\bar{y}$ that occur in this time interval in the absence of treatment. This plan is discussed in the following section.

7.4   THE SINGLE GROUP: SERIES OF MEASUREMENTS BEFORE AND AFTER

We therefore need a presumption that $\bar{y}_a - \bar{y}_b$ is of a size or direction that calls for explanation by major causes peculiar to the interval in question, of which the treatment may be one. The nature of alternative contributing causes will, of course, depend greatly on the type of study and the time interval elapsed between the pretreatment and the posttreatment measurements. The following are some examples.

Campbell and Stanley report Collier's (1944) study (actually a two-group study) of the effect on students of reading Nazi propaganda in 1940. The fall of France, which occurred during the time interval, may have had a major effect on attitude changes.

With an economic program intended to improve people's well-being, their final reports on the program may differ in a period of generally rising prosperity from those in a period of declining prosperity.

In evaluation of a program intended to increase worker satisfaction with working conditions in an agency, the questions asked may suggest to the worker the type of response acceptable to the management, which may, in turn, affect some reported changes in attitudes. Ingenuity and effort to ensure that reports are both anonymous and believed to be anonymous and skill in selection and wording of questions may help.

One precaution, especially when considerable time elapses between the pretreatment and the posttreatment measurements, is to keep the definition of measurement used for the standard and that for the level of measurement of response the same, particularly where human judgment or relations keep us involved in the measuring process.

Possible effects of the pretreatment measurements or of the knowledge that such measurements are being made, must also be considered. This issue is familiar in studies of teaching methods or of new devices in learning, where the response measurements are some kind of examination and the same persons are measured before and after. Quite apart from any treatment effect, students may perform better in a second posttreatment test because they are more mature or more familiar with the type of test; they could perform somewhat worse in the test if they have become bored or antagonistic. In instructional material to farmers on farm-management practices where the measurements are made by a skilled economist, a comprehensive initial questionnaire may suggest ideas to some farmers that induce changes quite apart from those at which the instructional material is directed.

With regard to the knowledge that pretreatment measurements would be made, Emmett (1966) reports an effect on the measures rather than on the subjects. The study involved the effect of radio programs in London which encouraged parents to bring their children to doctors or clinics for the standard children's inoculations. The numbers of inoculations per week during the two weeks before the radio programs and the three weeks after the programs were aired were as follows:

	Inoculations Per Week Two Weeks Before	Three Weeks After
118 general practitioners	209	130
5 clinics	85	79

A factor here was apparently that the doctors and clinics, alerted that the radio programs would take place, made special efforts in the two weeks before the programs to get children who were due for inoculations to come in so that the doctors or clinics would be free to handle the anticipated rush of patients into the office after the radio program.

If nonoverlapping samples are drawn for the pretreatment and posttreatment measurements, then distortion from the pretreatment measurement is unlikely to be large. The elapsed time between pre- and post-measurements or other considerations may dictate that independent measurements be used. For example, in a fluoridation study, we would probably want the data to refer to children of specific ages, who would be different individuals pretreatment and posttreatment. If there is worry about the effect of pretreatment measurement in the case of a treatment of short duration to which all are exposed, it might sometimes be feasible to draw initially a sample of size $2n$—twice the planned size. This sample is then divided into random halves, one to be pretreatment measured and one posttreatment measured, or even into n matched pairs, of which one member at random is pretreatment measured and one posttreatment measured. The primary penalty from using independent samples is a decreased precision in $\bar{y}_a - \bar{y}_b$. Its variance under the simplest model is $2\sigma^2(1 - \rho)/n$ with identical subjects, where ρ is the intrasubject correlation, versus $2\sigma^2/n$ or $2\sigma^2(1 - \rho_m)/n$, where ρ_m is the correlation between members of a matched pair.

If the cost of measurement is not excessive, another possibility with a treatment to which all are exposed is as follows. Having created the two groups of n random or paired samples, measure the first n samples before treatment and all $2n$ samples after treatment. Let the means of the three groups of n measurements be $\bar{y}_{1b}$, $\bar{y}_{1a}$ and $\bar{y}_{2a}$. The comparison $\bar{y}_{1a} - \bar{y}_{2a}$ provides a test of the effect of pretreatment measurement. If an effect is detected, the effect of the treatment is estimated by the comparison $\bar{y}_{2a} - \bar{y}_{1b}$. If no pretreatment measurement effect is found, the mean $(\bar{y}_{2a} + \bar{y}_{1a})/2$ may be compared with $\bar{y}_{1b}$ to estimate the treatment effect. Apart from other possible sources of bias mentioned in this section, these estimates are subject to some bias because the type of estimate used is determined by the result of a preliminary test of significance, but this bias should be small in large samples. When no effect of pretreatment measurement exists, it is theoretically possible to obtain a more-precise estimate of the treatment effect by using a weighted, instead of an unweighted, mean of $\bar{y}_{1a}$ and $\bar{y}_{2a}$ to take fullest advantage of the correlation between $\bar{y}_{1b}$ and $\bar{y}_{1a}$ which refer to the same subjects. If ρ were known, under a constant variance σ^2, the best weighting would be $[(1 + \rho)\bar{y}_{1a} + (1 - \rho)\bar{y}_{2a}]/2$. But the potential gain over equal weighting does not become worthwhile until ρ reaches 0.6, the variance of the estimated treatment effect being $(3 + \rho)(1 - \rho)\sigma^2/2n$ with optimum weighting and $(3 - 2\rho)\sigma^2/2n$ with equal weighting.

To summarize, avoid the single-group before–after study unless nothing better is practicable. By itself, the difference $\bar{y}_a - \bar{y}_b$ provides no basis for speculation about a treatment effect, unless there are strong grounds for concluding that this difference is of a size that would not have occurred in the time interval involved in the absence of a treatment effect or other major cause. In attempting conclusions, the investigator has the responsibility of using imagination and help from colleagues in listing alternative major contributing causes. For each such cause the investigator should consider any evidence that assists a judgment about the size and direction of the resultant effect. Discussion of these alternative causes and the investigator's judgment about the biases that they create should be included in the report of study results.

Given a time series of measurements at intervals τ both before and after the introduction of the treatment, we are in a better position to judge whether an unusual change in the time series coincided with the introduction of the treatment. Campbell and Stanley (1963/1966) illustrate a variety of situations that may occur with four pretreatment and four posttreatment measurements. Some situations strongly suggest an unusual change in level or direction; in some the verdict on this point is dubious, while in others a glance shows no sign of anything unusual.

Thus in Figure 7.4.1, situation A indicates an unusual rise in level that persists after introduction of the treatment, while situation B not only

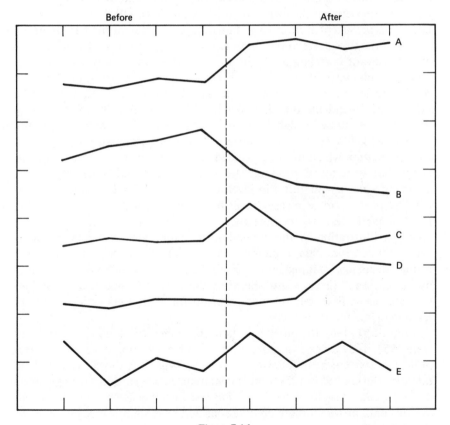

Figure 7.4.1

indicates a shift in level, but hints at a reversal of the time trend. In situation C, the shift in level lasts for only one time interval, which would be consistent with a treatment effect that was predicted beforehand to be of short duration. Situation D is more ambiguous, a shift occurring two intervals after introduction, which might either represent a delayed effect or some later causal force having nothing to do with the treatment. Situation E is one of many indicating nothing unusual occurring in the key interval.

It would help if visual impression of the before and after measurements could be made more objective by a valid test of significance of the postulated effect of the treatment. In short series the possibilities are limited to the simplest situations. In situation A, if successive observations in the time series could be regarded as independent, a standard t test of $\bar{y}_a - \bar{y}_b$ with 6 degrees of freedom could be used. However, in most time series, successive observations are correlated, the correlations depending on the time interval between the observations. Box and Tiao (1965) have given two tests for a persistent shift in level, based on two different models about the nature of the time correlations, but these tests require knowledge of the sizes of the relevant correlations. With a time series that clearly has rising or falling trends, tests of significance based on linear-regression models (e.g., changes in levels, slopes, or both) are easily constructed if residuals can be assumed independent, the chief trouble being the paucity of degrees of freedom likely to be available for estimating the residual variance. I do not know any likely tests related to linear trends when residuals are correlated. One temptation which must be avoided is to make the postulated treatment effect that is tested depend on the appearance of the time series (e.g., assuming a persistent shift δ in level in A in Figure 7.4.1, a shift δ only in the first posttreatment measurement in C, and delayed-treatment effect in D.) This tactic destroys the objectivity of the test.

As an illustration of the methodological approach in handling "interrupted" time-series data, Campbell and Ross (1968) consider an evaluation of the effects, on fatalities in motor accidents, of a crackdown on speeding by suspension of licenses, introduced by the state of Connecticut in December 1955, using five "before" measurements for 1951–1955 and four "after" measurements for 1956–1959.

From 1951–1955 the number of fatalities indicated an upward trend and from 1956–1959 a downward trend, the graph suggesting a definite reversal in the fatality-rate trend following the crackdown. Year-to-year variations in fatalities during 1951–1959 were substantial, however, and an application by G. V. Glass (1968) of the Box–Tiao test for a persistent shift in level to *monthly* data provided more observations and gave significance just short of the 10% level.

If we accept this as partial evidence for an unusual force at work beginning in 1956, Campbell and Ross proceed to consider the crackdown and possible alternative contributors as causal factors. With regard to the crackdown treatment, they plot, from 1951 to 1959, the suspensions of licenses for speeding as a percentage of all license suspensions. This graph shows a marked rise in license suspensions to a new level in 1956, as evidence that the crackdown was actually applied. Similarly, a graph of the percentage of speeding violations to all traffic violations showed a marked drop from 1956 onward.

Among *possible* alternative contributors as causal factors, Campbell and Ross consider a dramatic improvement in the safety features of cars and a decrease in the amount of hazardous driving conditions, concluding that neither of these constitutes a plausible rival hypothesis in this case. Another source to be considered is any possible consequence of the 1955 pretreatment measurement. As it happened there was about a 35% increase in deaths from 1954 to 1955, the latter figure being the highest ever in Connecticut. There may have been two effects of the initiation of the crackdown following this maximum. First, publicity given to the 1955 increase in deaths may have induced a period of more careful, defensive driving. Second, in a time series in which there are annual ups and downs due to a multiplicity of influences, an exceptionally high value, as was 1955, is likely to be due in part to an unusual combination of upward influences in 1955 and therefore to be followed by several lower values. As often happens in this type of study, it is difficult to estimate whether either of these effects was substantial. Another source to be investigated is any change in the standard of measurement or record keeping. This does not seem to be a contributor in this study, but can be major in some evaluations of new administrative practices, where a marked change in record-keeping habits, accuracy, and detail may be introduced as a part of the new program.

Even if there is no *planned* comparison group from which the treatment in question was absent, it is worthwhile to consider whether something can be learned from data available for other populations not strictly comparable. Campbell and Ross use this idea to compare traffic deaths per 100,000 persons during 1951–1959 for Connecticut and the nearby states of New York, New Jersey, Rhode Island, and Massachusetts. These states also show substantial year-to-year variations, but overall comparison of the post-1956 with a pre-1956 record of fatalities is more favorable in Connecticut than in the other states.

The concluding judgment of Campbell and Ross is "As to fatalities, we find a sustained trend toward reduction, but no unequivocal proof that they were due to the crackdown. The likelihood that the very high prior rate instigated the crackdown seriously complicates the inference." These statements illustrate the type of conclusion that an observational study permits —one that nearly always involves an element of the investigator's judgment.

The objective of this study was to measure the effect of the crackdown on a single response variable—number of traffic deaths. In the broader aspects of program evaluation the investigator must choose response variables that reflect not only the intended effects of the program, but any other effects that he judges important. In this connection Campbell and Ross note, incidentally, from an examination of the 1951–1959 figures, two other indicated changes in 1956 that *may* represent less desirable effects of the crackdown. Relative to suspensions there was a marked increase in the proportion of arrests while driving with a suspended license, and also in the percent of speeding violation charges judged not guilty in the courts.

In a situation somewhat analogous to the time-series comparison, it may occasionally be possible to obtain some independent comparisons with and without the treatment in question. An example is the attempt to estimate how much the advertised appearance of a famous pitcher A adds to the attendance at a baseball game. In the professional leagues, two teams normally play on three or four successive days in the same park. The approach used is to compare the attendance on the days when A is pitching with the attendance on neighboring days that are similar as to weather, time of day or night, day of the week, and so forth. The power of the pitcher to draw a crowd for the other team is a confounding factor, but repetitions are obtained because pitcher A appears in numerous parks in different cities against different teams. The type of approach needed is similar to that in the Connecticut crackdown.

Although Cochran planned several more sections for this chapter, his writing stopped at this point. [The editors]

REFERENCES

Box, G. E. P. and G. C. Tiao (1965). A change in level of a non-stationary time series. *Biometrika*, **52**, 181–192.

Campbell, D. T. and H. L. Ross (1968). The Connecticut crackdown on speeding: Time series data in quasi-experimental analysis. *Law and Society Review*, 3(1), 33–53.

Campbell, D. T. and J. C. Stanley (1963). Experimental and quasi-experimental designs for research on teaching, in N. L. Gage, Ed. *Handbook of Research on Teaching*. Rand McNally, Chicago. (Also published as *Experimental and Quasi-Experimental Designs for Research*. Rand McNally, Chicago, 1966.)

Collier, R. M. (1944). The effect of propaganda upon attitude following a critical examination of the propaganda itself. *J. Soc. Psychol.*, **20**, 3–17.

Emmett, B. P. (1966). The design of investigations into the effects of radio and television programmes and other mass communications (with Discussion). *J. Roy. Statist. Soc. Ser. A*, **129**, 26–59.

Glass, G. V. (1968). Analysis of data on the Connecticut speeding crackdown as a time-series quasi-experiment. *Law and Society Review*, **3**(1), 55–76.

Additional Reference

Had he been aware of this reference we believe that Cochran would have added it to the list for this chapter. [The editors]

Cook, T. D. and D. T. Campbell (1979). *Quasi-Experimentation: Design and Analysis Issues for Field Settings*. Rand McNally, Chicago.

Index

Page numbers in *italics* refer to full bibliographic references.

Adjustment *vs.* matching, 102–103
  in experiments, 119
  in observational studies, 121
Adjustment by regression, 75
Agriculture, gains in, 3
Analysis of covariance, 10, 75. *See also*
  Covariance
Analysis of variance, 75
Analytical surveys, 2, 3, 5
Announcement, pre-treatment, effect of, 133

Bartlett, M. S. 123, *128*
Beets, sugar, 69
Belson, W. A., 112, 113, *128*
Bias:
  adjustment for, 103
  appraising, 12
  from confounding, 74
  consistency of, 43
  control of, 8
  in controls, 42
  of differences, 17, 90
    regression models for, 90
  detection of, 37
  estimating, 12, 13
  everpresent, 26
  examples of, 24
  limits for, 26
  model for, 17
  from nonresponse, 66
  in observatational studies, 12
  in population, 22, 23
  removal of, 28
  sources of, 8
Bias effect:
  adjustments for, 112
  on confidence level, 26, 27
  on significance level, 58
  test of significance for, 28
Bias reduction:
  caliper matching, 88
  matching, 85
  mean matching, 90
  nearest available matching, 89
Billewicz, W. Z., 78, 82, 96, *100*, 111, 117,
  128
Blindness, 8
  of judges, 45
Blocking, 9, 75
Bombing raids, effect on German factories, 44
Boruch, R.F., 121, *129*
Box, G. E. P., 136, *138*
Brain-damaged boys, 77–78
Buck, A. A., 33, *49*

Caliper matching, 79, 88
Call-back, 67
Campbell, D.T., 121, *128*, 130, 132, 135–138,
  138, 139
Campbell, E.Q., *14*
Cause, 1, 3
Chi-squared tests, 108
Classes:
  optimum, 38

141